Undergraduate Lecture Notes in Physics

Series Editors

Neil Ashby, University of Colorado, Boulder, CO, USA

William Brantley, Department of Physics, Furman University, Greenville, SC, USA

Matthew Deady, Physics Program, Bard College, Annandale-on-Hudson, NY, USA

Michael Fowler, Department of Physics, University of Virginia, Charlottesville, VA, USA

Morten Hjorth-Jensen, Department of Physics, University of Oslo, Oslo, Norway

Michael Inglis, Department of Physical Sciences, SUNY Suffolk County Community College, Selden, NY, USA

Undergraduate Lecture Notes in Physics (ULNP) publishes authoritative texts covering topics throughout pure and applied physics. Each title in the series is suitable as a basis for undergraduate instruction, typically containing practice problems, worked examples, chapter summaries, and suggestions for further reading.

ULNP titles must provide at least one of the following:

- An exceptionally clear and concise treatment of a standard undergraduate subject.
- A solid undergraduate-level introduction to a graduate, advanced, or non-standard subject.
- A novel perspective or an unusual approach to teaching a subject.

ULNP especially encourages new, original, and idiosyncratic approaches to physics teaching at the undergraduate level.

The purpose of ULNP is to provide intriguing, absorbing books that will continue to be the reader's preferred reference throughout their academic career.

More information about this series at http://www.springer.com/series/8917

Giulio Magli

Archaeoastronomy

Introduction to the Science of Stars and Stones

Second Edition

 Springer

Giulio Magli
Department of Mathematics
Politecnico di Milano
Milan, Italy

ISSN 2192-4791 ISSN 2192-4805 (electronic)
Undergraduate Lecture Notes in Physics
ISBN 978-3-030-45146-2 ISBN 978-3-030-45147-9 (eBook)
https://doi.org/10.1007/978-3-030-45147-9

All illustrations courtesy Emanuela Franzoni
If not otherwise stated, all photographs courtesy Giulio Magli

This Springer imprint is published by the registered company Springer Nature Switzerland AG
The registered company address is: Gewerbestrasse 11, 6330 Cham, Switzerland

Accessing Augmented Reality Augmented reality is an integral part of the second edition of "Archaeoastronomy. Introduction to the Science of Stars and Stones": to enhance your reading experience, hold your mobile device over illustrations or photos in the book and spot supplemental multimedia content, such as short videos which explain the phenomena or 3D buildings you can interact with.

Installing the "ArchaeoastronomyAR" App

To unlock these special features, we recommend downloading the free "ArchaeoAR" Augmented Reality App for smartphones and tablet PCs from the app stores:

Consult the list of recommended devices that are enabled at this page: https://library.vuforia.com/articles/Solution/vuforia-fusion-supported-devices.html. This list does not represent all of the devices that can support Vuforia Engine SDK vision-only and Fusion features. Additional platform supported Android devices that are supported via Vuforia Fusion can be found here: https://developers.google.com/ar/discover/supported-devices.

Using the "ArchaeoastronomyAR" App The app allows you to scan the pictures of the book showing this ⊙ icon: related videos will appear on your device's screen just over them. Look for the arrow icon throughout the book to spot the pictures which have more elements to discover.

You can always view 3D monuments by accessing them directly from your handheld device, using the app's menu.

From the app's menu you can:

1. Frame a picture showing the ⊙ icon, to play a video.
2. Visualise the following 3D buildings on your device's screen.

To visualise a 3D monument, hold your smartphone or tablet parallel to a horizontal surface whose colour is not uniform and click on the tracker once the grid appears. You can interact with the monuments, thanks to some specific features which will allow you to understand the archaeoastronomical phenomena.

Contents

Part I   Methods

1   Astronomy with the Naked Eye 3
   1.1   The Celestial Coordinates 3
   1.2   The Apparent Motion of the Sun 6
   1.3   The Apparent Motion of the Stars 11
   1.4   Constellations 13
   1.5   The Milky Way 16
   1.6   Precession 16
   1.7   The Apparent Motion of the Moon 19
   1.8   The Apparent Motion of the Planets 22
   1.9   Calendars 25
   1.10 The Observation of Celestial Bodies on the Horizon 27
   References 28

2   Acquiring Data 31
   2.1   Archaeoastronomy Fieldwork 31
   2.2   The Magnetic Compass 33
   2.3   The Clinometer 35
   2.4   The Theodolite 36
   2.5   The Global Positioning System 37
   2.6   Virtual Globe Software 39
   References 41

3   Data Analysis 43
   3.1   Reconstructing the Ancient Sky 43
   3.2   The Reconstruction of Visual Alignments and the Horizon
         Formula 44
   3.3   Graphical Tools 46
   3.4   Statistical Tools 48
   References 51

Part II Ideas

4   Astronomy and Architecture at the Roots of Civilization   55
    4.1   From Homo Sapiens to Homo Sapiens   55
    4.2   From Hunters-Gatherers to Herders-Peasants   61
    4.3   The Birth of Monumental Architecture   63
    4.4   The Birth of Astronomically Anchored Monumental
          Architecture   70
    References   75

5   Astronomy, Power, and Landscapes of Power   77
    5.1   Sky and Cosmos   77
    5.2   Cosmos and Afterworld   79
    5.3   Mastering the Cosmos   82
    5.4   Cosmic Machines   85
    5.5   Cosmic Landscapes   88
    References   100

6   The Scientific Foundations of Archaeoastronomy   101
    6.1   Archaeoastronomy as a Cognitive Science   101
    6.2   Archaeoastronomy as an Exact Science   106
    6.3   Archaeoastronomy and Unwritten Sources   110
    6.4   Archaeoastronomy and Reverse Engineering   116
    6.5   Towards Archaeology of the Cosmic Landscape   119
    References   121

Part III Places

7   Megalithic Cultures of the Mediterranean   125
    7.1   Stonehenge and Its Landscape   125
    7.2   Newgrange and the Bend of the Boinne   136
    7.3   The Sleeping Giant   143
    7.4   Taulas and Stars   147
    References   152

8   Ancient Egypt   155
    8.1   A Seat Among the Imperishable   155
    8.2   The Horizon of Khufu   164
    8.3   The Giza Written Landscape   169
    8.4   The Sun in the Temples   173
    References   179

9   Pre-Columbian Cultures   183
    9.1   The Maya at Uxmal: The Governor's Palace   183
    9.2   The Serpent on the Pyramid: Chichen Itza'   189

9.3   Going Where the Sun Turns Back   200
9.4   Pillars of the Sun   207
References   216

10 The Classical World   219
10.1   Houses of the Gods   219
10.2   The City of the Lion   224
10.3   A Comet and a Capricorn: Augustus' Power from the Stars . . .   229
10.4   Astronomy and Empire at the Pantheon   237
References   242

11 Asian Cultures   245
11.1   The Terracotta Army   245
11.2   The Pyramids of Ancient China   247
11.3   Between Sun and Waters: Angkor Wat and the Khmer
        State-Temples   251
11.4   Perennial Hierophanies in the Khmer Heartland   254
11.5   The "Sun Path" at Borobudur   257
References   258

Exercises   261

Introduction

It was 21 December 1969. A few minutes before dawn in Newgrange, Ireland.

Archaeologist Michael J. O'Kelly carefully checked that his latest discovery, a narrow opening located directly above the entrance to the tomb, was unobstructed by any pieces of wood or debris. Then, he closed the door, walked up the corridor to the end room and sat down. He waited. Shortly afterwards, the Sun appeared on the horizon, filtered through the window, reached the chamber and cast a ray. In this way, an appointment between stars and stones that had been set up 5000 years previously was finally attended again. And a new science, archaeoastronomy, was officially born.

Since then, archaeoastronomy has undergone a long and sometimes arduous development, gradually leading up to the comprehensive, well-grounded scientific discipline that it is today. However, all the ingredients were already in place at that moment: a monument built by knowledgeable architects, a shining star, and a scientist, eager to understand how and why those ancient architects connected their building with the sky.

This book is intended as an up-to-date, easy-to-follow presentation of the subject. As a second, but equally important aim, it also intends to arm the reader with the necessary set of instruments that will enable him to distinguish between serious archaeoastronomical research and the vast amount of pseudo-archaeological claptrap masquerading as archaeoastronomy which is readily available on the Web and in bookshops. The volume is self-contained and can therefore be approached by any curious person, from the archaeologist or astronomer interested in the applications of this discipline to the general reader. However, it has been designed basically as an undergraduate textbook and is the fruit of my eight years experience of teaching the archaeoastronomy course at the Faculty of Civil Architecture in the Politecnico of Milan (indeed it is the only course of its kind offered in an Italian university). And of course, any academic book worth its salt would not be complete without a section of exercises at the end.

The material is organised as follows. I have with some reluctance included the concepts that are essential for grasping technical aspects of the subject—for example, in learning how to carry out the archaeoastronomical analysis of a site—in Part I. The reason is that any self-respecting course in archaeoastronomy must first deal with at least the basics of such notions. As a result, from the very first pages the reader will be faced with technical material, such as astronomy with the naked eye, positional astronomy and data analysis. This part may be potentially off-putting for even the most ardent reader, and my suggestion is that the author's order need not be strictly respected. One possibility may be to start studying only Sects. 1.1–1.6 of Chap. 1 (basic astronomy with the naked eye of the Sun and the stars, and precession) and then proceed, considering the rest of Chap. 1 as a handy reference guide if a particular concept needs to be explored. Chapters 2 and 3 can then be browsed through briefly, and the bulk of Part I left to be studied at the end, in order to have a complete overview and tackle the final exercises and/or fieldwork. A similar line could be followed (and it is actually followed by myself) in teaching the same material.

Part II is divided into three chapters. Chapter 4 discusses at some length the birth of the relationship between astronomy and architecture and can be omitted at a first reading. It shows how the two have developed side by side since ancient times and seeks to demonstrate that they actually started together. The reasons behind this connection are explored at length in Chap. 5, which is devoted to the nexus between religion, power, astronomy and architecture. Key concepts such as hierophany and sacred (or cosmic) landscape are introduced and discussed against the historical and anthropological background of archaeoastronomy. Chapter 6 then explores the scientific foundations of archaeoastronomy as well as its relationship with the humanities. The goal is essentially to emphasise the ways in which this discipline can be used as a tool in understanding some significant aspects of the past civilisations, but also to warn against the dangers of indiscriminate or possibly naive use.

Part III is devoted to the detailed description of a dozen or so significant places. The choice of the sites has been made with the objective of giving a sample of achievements made but also of explaining some of the problems inherent in this discipline; the list contains several well-known, spectacular monuments and landscapes but also less famous ones which are, nevertheless, fascinating and instructive from various viewpoints. There is of course no aim at completeness, and I apologise in advance to any reader who might be disappointed at not finding a particular country or site mentioned. However, after studying this book the reader will be able to read the relevant literature on any place or site that has piqued his interest and seek it out, both virtually and in reality. Further, most of the places that I have only touched on briefly in the book are discussed—concisely but effectively—in the entries of the *Handbook of Archaeoastronomy and Ethnoastronomy*, edited by Clive Ruggles (2015).

As mentioned, the book ends with a series of exercises which the reader can attempt using free astronomical and virtual globe software. Of course, there is no substitute for field experience, which is recommended to all readers whenever they have the opportunity. However, this software allows us to simulate fieldwork on our own PCs, and this greatly facilitates teaching—and learning—archaeoastronomy. Besides, leaving aside the Sun, the appointments fixed with the other stars are no longer valid today, due to the phenomenon called precession, and therefore have to be recreated with a computer anyway.

This new (2020) edition of the book features a brand new chapter devoted to Asia. Further, it implements augmented reality to enhance the reading and learning experience. Holding indeed a mobile device over many (more than 50) of the illustrations or photos in the book, short videos will be displayed taken from the MOOC (massive open online course) developed at Politecnico di Milano and avaliable in full on the Web platforms Polimi Open Knowledge (www.pok.polimi.it) and Coursera (www.coursera.org/learn/archaeoastronomy).

Well, without further ado, let me lead you into this book by borrowing two verses from a Italian poet whose identity, I believe, is not infernally difficult to guess. The verses, supposedly pronounced at dawn a few days after the spring equinox of 1301, read: *Ma seguimi oramai che 'l gir mi piace/ché i Pesci guizzan su per l'orizzonta*, that is (author's translation): *But now follow me, because I would like to go/since the Fishes are already quivering on the horizon.*

Part I
Methods

Chapter 1
Astronomy with the Naked Eye

1.1 The Celestial Coordinates

Today we explore the sky with powerful instruments and we know that there are millions of galaxies in the universe, each containing billions of stars. We are accustomed to thinking about the universe from an "unprivileged" point of view. We know, moreover, that the Sun is only one among billions of stars, and a pretty bog standard one at that, and that it occupies a relatively insignificant position in one of the branches—itself quite bog standard—of one of those galaxies: the Milky Way.

Many ancient cultures acquired a profound knowledge of the sky independently, but the celestial bodies available for their observations were only those visible with the naked human eye. Furthermore, their point of view was always that of an observer who is located on an object—the Earth—which is subject to a complex series of motions of its own. The continual changes in the sky observed by them were, therefore, for the most part, due to *apparent* movements. In spite of this, many ancient astronomers devised ingenious methods of observation and made highly accurate measurements, which were reflected and exploited in the projects of the monumental architecture of their historical epochs. Thus if we want to understand these (in some cases fundamental) aspects of architecture, we must first study and understand astronomy as they did: with the naked eye, and from the Earth's surface.

The centre of mass of our planet moves on an ellipse, with the Sun located at one focus. The plane which contains the Sun and the orbit of the Earth is called the *Ecliptic*. The Earth's axis is not perpendicular to the Ecliptic, but is (today) inclined by $23°30'$; this angle is called *obliquity* and denoted with the Greek letter ε. Since the Earth's axis wobbles slightly, the value of the obliquity is not strictly constant. It varies between $25°$ and $22°$, currently diminishing at a rate of about $48''$ per century, so that it was $30'$ greater 4000 years ago (Aveni 2001).

© Springer Nature Switzerland AG 2020
G. Magli, *Archaeoastronomy*, Undergraduate Lecture Notes in Physics,
https://doi.org/10.1007/978-3-030-45147-9_1

Consider now a fixed observer on the Earth's surface. The observer sees all heavenly bodies move as a consequence of his own movements, in particular, the daily terrestrial rotation. The observer's viewpoint is a classic example of a very *poor* frame of reference in Mechanics, since the complicated amalgam of our planet's movements give rise to quite complicated relative dynamics (a much better viewpoint would, of course, be from a spacecraft that is stationary in relation to the solar system).

To become acquainted with a reference system, one needs a system of coordinates, that is, a set of numbers which allows one to identify unambiguously the position of any object. The observer on the Earth sees the sky as a spherical surface, the celestial sphere. In other words, he is generally unable to discern if, for instance, two stars which appear to be close in the sky are really close in the radial direction or not. Since to fix a point on a sphere only two numbers are needed (for instance, latitude and longitude on the Earth), each system of celestial coordinates will associate two numerical values to any object seen in the sky at a certain time. There are two systems of coordinates which are commonly used (and useful) in archaeoastronomy. The simpler one is the azimuth-altitude system; to visualise it, imagine first projecting the course of the Earth axis onto the celestial sphere. There will be a ideal point of intersection in the sky which we shall call the celestial pole (north or south depending on the hemisphere of the observer; but to simplify matters, unless otherwise specified, I shall refer to the north pole from now on). This ideal point is of fundamental importance; indeed, since our own reference system is rotating around the Earth's axis, we see everything in the sky rotating around the pole. Now, we should imagine lowering the perpendicular to the horizon from the celestial pole. In this way, we can identify a point on the horizon and a direction on the ground towards it: these are known as *geographical north* and the *meridian*. Given a point S in the sky, whose coordinates we wish to find, we now imagine tracing the vertical plane which passes through this point. This plane intersects the horizon of the observer at a point, say S*; the *azimuth* A is the angle between north and the point S* on the horizon, counting positively from north to east (in other words, clockwise), and the *altitude* a is the angle measured on the vertical circle from S* to S (Fig. 1.1). In particular, the altitude reached by a star when it passes the celestial meridian, that is, the ideal projection of the observer's meridian into the sky, is called *culmination*. Of course, the celestial pole is located on the meridian; it thus has a constant azimuth (zero) and a constant altitude which, as is easy to see, equals the latitude of the observer.

The second system of coordinates we need to use does not depend on the choice of our terrestrial observer. This system is easily understood if one imagines measuring the latitude and longitude of a point *on* the celestial sphere. The angular distance of the point from the celestial equator gives an angle called *declination*; the analogue of the longitude is called *right ascension*, and it is conventionally measured from a point (the vernal equinox) along the celestial equator to the maximal circle passing through the pole and the point (Fig. 1.2).

Fig. 1.1 The altitude/
azimuth coordinate system.
Both coordinates of a *star*
change continuously ⊙

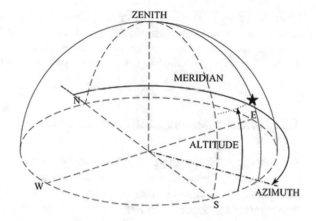

Fig. 1.2 The right ascension/
declination coordinate system.
Stars move on *circles* of fixed
declination δ ⊙

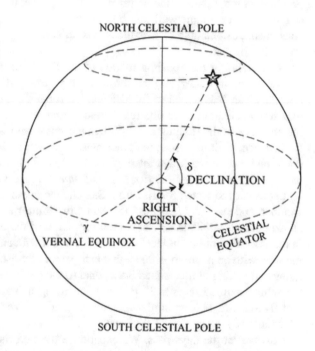

The right ascension-declination system has, for the archaeoastronomer, two
important advantages. First of all, to know where an object was located in the sky in
a certain period in the past, only one number is needed, namely, declination.
Furthermore, declination is independent from the latitude of the observer, while the
azimuth is. It is, therefore, important to be able to treat experimental data (usually
acquired using the azimuth-altitude system) in terms of declinations. This is done
using mathematical formulae which connect the two systems; one very useful

formula for us is the one giving the (sinus of) the declination δ of a star as a function of azimuth and altitude, once the latitude λ of the observer is known:

$$\sin \delta = \sin a \sin \lambda + \cos a \cos A \cos \lambda.$$

1.2   The Apparent Motion of the Sun

A stable star, of which there are billions in the universe, can be broadly described as a huge sphere of plasma, that is, a (globally neutral) medium where all the particles are unbound and charged. Gravity tends to shrink (the technical term is collapse) the object, but thermonuclear reactions take place, which in turn produce a radiation pressure which balances the contraction. As a consequence, a standard star is in a stable state that might well last billions of years, until its nuclear fuel has been consumed. An example of a stable star is, unsurprisingly, the one that assures life on Earth, the Sun.

Understanding the apparent motion of the Sun is of fundamental importance in archaeoastronomy. Let us, then, proceed very slowly. For a reason that will soon become clear, it is useful to fix ideas on the movement of the Sun by considering two different examples, at different latitudes. One such latitude must be greater than the obliquity of the Ecliptic ε (=23°30' in present times), so we could have our base, for instance, at Stonehenge; the other must be lesser, so for instance in this case we could be located in a Maya town of the Yucatan. In both cases, *we assume for the moment that our horizon is perfectly flat and disregard any atmospheric effect*.

Let us start to watch Sunrise from Stonehenge on an arbitrary day, say the 1st of October, and see what happens. The Sun's first limb rises on the eastern horizon at an azimuth of $\sim 95°$, that is an azimuth 5° south of the east. Our star makes an arc in the sky culminating, at local noon, at due south with an height of 32°, then sets in the west with an azimuth $\sim 265°$, that is 5° south of west. The segment between the rising and setting points is thus orthogonal to the meridian or, to put it another way: the setting azimuth equals 360° minus the rising azimuth, something which is in fact always true, at least in ideal conditions (zero horizon, zero atmospheric effects, as mentioned) every day of the year.

Now we let the days pass. We experience the Sun rising more and more to the south of east (so the rising azimuth increases) and culminating lower and lower (so culmination altitude decreases). In fact we might even start to panic that the Sun will eventually fade away at due south and vanish. But the process also slows down, so that the rising azimuths on successive days finally differ by such a small amount that it becomes very difficult to distinguish between different positions on the horizon. Gradually we will come to realise that the movement of the rising Sun, seen as a point on the horizon in successive days, has inverted its direction. The southernmost azimuth was reached on 21 December, the day we call the winter solstice. So we are reassured and can resume following the Sun rising on the horizon. The rising azimuth decreases, the culmination angle increases, and the Sun again takes up its previous positions on the horizon on dates which are symmetrical in comparison with the previous ones. Finally, the Sun rises at due east, thus assuming a declination equal to zero: the (astronomical) equinox. Then the rising azimuth is seen to decrease more and more. Once again, however, we also experience a slowing down until we reach the point where we are unable to distinguish different positions from each other, and finally an inversion occurs on the 21 June, the summer solstice. On this day, the rising azimuth is symmetrical, with respect to due east, to that of the winter solstice, and the culmination of the Sun is the highest possible, ~61°30' at Stonehenge. After the solstice, the Sun will retrace its steps, moving towards the Autumn equinox (Fig. 1.3).

To sum up, there are 4 days which can naturally be singled out: the two solstices and the two equinoxes. A natural way of dividing the year would thus be into four quarters; but due to the fact that the Earth's orbit is elliptical, the equinoxes are not separated precisely by an equal number of days from the solstices. As a result, the astronomical definition of equinox can be misleading in the historical sense, since it is more likely that ancient civilizations were interested in marking quarters days of the year, i.e. midpoints in time between the solstices, dividing the days between the two solstices into quarters of 91 days each (plus one day). These intermediate days fall very close, but not exactly on, to the equinoxes (they fall on March 23 and

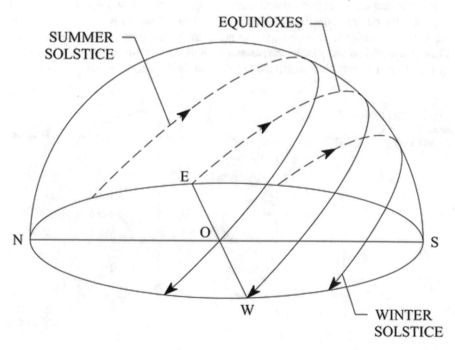

Fig. 1.3 The path of the sun during the year, viewed by an observer at a latitude above the tropic, with a flat horizon ⊙

September 21, ±1 day, in relation to equinoxes on March 20 and September 22/23). In any case, we must not make the mistake of thinking that only solstices and possibly equinoxes (or quarter days) were considered important in antiquity: in other words solar archaeoastronomy is *not* just "talking about solstices and equinoxes", since other solar dates might have been equally, if not more, important for cultural reasons (e.g. the date of foundation of Rome in the Roman world, or the zenith passages of the Sun in tropical zones, as defined below). With regard to solstices, it is worth stressing that the Sun never *stops* at the solstices, in other words, the solstice azimuth is strictly reached only once; it is, however, difficult to distinguish the azimuths between—say—the solstice and three days before or after, because the rate of change of the azimuth is very low (in the order of 1 prime of arc per day). On the contrary, this rate increases when the Sun is closer to the equinoxes, where the difference in two subsequent positions of the Sun at the horizon reaches 25′, that is, almost a solar diameter (the solar diameter is 30′ as seen from the Earth).

A similar behaviour in terms of declination corresponds to this sinusoidal change in the azimuth. In fact a useful formula for the change in declination of the Sun is the following:

$$\sin \delta = \sin \varepsilon \cos w$$

where w = 0.986 n and n is the number of days that have elapsed since the June solstice expressed in degrees (so that for n = 0, cos w = 1 and δ = ε). When we approach the equinox, declination goes to zero (and indeed for n ∼ 90, cos w ∼ 0) while for n = 180, cos w ∼ −1 and we are close to the opposite solstice (δ = −ε). Finally, a useful quantity is the measurement of the arc (symmetrical with respect to true east) of the possible rising points of the Sun in the course of the year, or *solar*

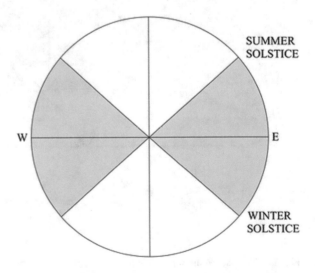

Fig. 1.4 The rising-setting azimuths of the sun corresponding to Fig. 1.3

Fig. 1.5 The path of the sun during the year, viewed by an observer at northern tropical latitude, with a flat horizon ⊙

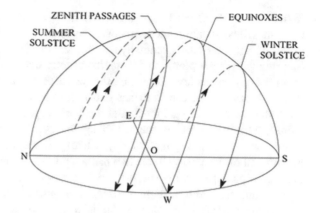

amplitude; it diminishes with decreasing latitude, being (with a flat horizon) in the order of 78° at Stonehenge, 64° in Rome, and 56° in Cairo (Fig. 1.4).

Correspondingly, the height of the Sun at culmination on the summer solstice *increases*, up to a latitude where our star passes directly overhead at the solstice. The corresponding parallels are called the Tropic of Cancer and the Tropic of Capricorn respectively; the reason is that these two zodiacal signs corresponded to the Sun at summer solstice in the two hemispheres about 2000 years ago (today, due to precession, it is not longer so, as will be clear in Sect. 1.6).

It is the existence of the Tropics that prompts the need to look carefully at the Sun's motion from tropical latitudes too—that is, latitudes between a tropic and the equator.

Let us now move to Chichen Itza' in the Yucatan, an ancient Mayan town where, as we shall see, astronomy played a significant role (Fig. 1.5).

The movement of the Sun *on the horizon* of Chichen Itza' is the same as at Stonehenge, the only difference being that the solar amplitude is smaller. The culmination of the Sun, on the other hand, is subject to a *qualitative* change. The

Fig. 1.6 The rising-setting azimuths of the sun corresponding to Fig. 1.5. The azimuths of the zenith passages are highlighted

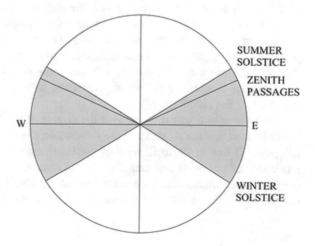

reason is that, as at Stonehenge, the nearer we get to the summer solstice, the higher the Sun will be at culmination; however, here on April 20, the height is already 80°, on May 18 it is 89°, and eventually the Sun reaches 90° on May 20. On this day, the Sun thus passes the zenith, that is, it sits directly overhead an observer at noon: the observer has no shadow. Logically, the day after the Sun should culminate higher than this, and this of course means that it culminates *to the north* of the observer. After the summer solstice, the Sun at the horizon retraces its steps, and therefore at culmination it approaches the zenith again; eventually we experience a second zenith passage on the symmetric day. Clearly, if we move to even lower latitudes, the interval of days between the two passages will increase, up to the equator where the zenith passages will occur at the equinoxes (Fig. 1.6).

In connection with the zenith passages, a curious fact can be mentioned. If we project the azimuth of the Sun on the days of zenith passages towards the western horizon, we get an azimuth to the south of west which corresponds to the setting of the Sun on two days between the autumn and the spring equinox. On these two days nothing especially noticeable happens in our hemisphere—in particular, the culmination of the Sun is *not* at the minimum, which always occurs at winter solstice. However, it is easy to see that, at midnight on these days, the Sun is passing at the zenith in the *opposite* hemisphere. For this reason, these two days are sometimes called *nadir passages*.

To conclude this discussion, we need to ask: what is the origin of this relatively complicated apparent motion of the Sun? Let us take for a moment the point of view of the observer outside the solar system, and look at the path our planet follows round the Sun. During one year the Earth (considered, for simplicity's sake, as a sphere, ignoring the slight flattening at the poles) moves on its orbit keeping constant the angle of the axis inclination in relation to the Ecliptic. This means that there will be, on the orbit, a point at which the tilt *towards* the Sun is maximum for, say, the northern hemisphere. This corresponds to the day of the summer solstice, as the angle formed by the the sun's rays (which are all to be considered as being parallel to each other) is maximal. The opposite situation will be the winter solstice (when it is the summer solstice for the southern hemisphere). Of all the intermediate situations, there are two points at which the direction of the tilt and the direction to the Sun are perpendicular: these are clearly the equinoxes. Finally, the Sun's rays can be perpendicular at noon on certain days (which are the zenith passages) only at points on the Earth which lie at latitudes lower than the obliquity; the Tropics therefore correspond to latitudes (north and south respectively) equal to ε, and this is why their location is not strictly constant in time but varies along with the variation of the obliquity.

Since the intensity of sunlight over the course of the year conditions the climate, it can be said that the presence of the seasons on the Earth (and probably of human life itself) is due to the inclination of the Earth axis. Yet defining—as we modern humans tend to do—the starting days of the seasons using equinoxes and solstices does not really make sense, since the path of the Sun in the sky in—say—the first decade of August is virtually the same as that of the second decade of May. The ancients knew this very well, so, for example, for the Romans the seasons usually had their midpoints, rather than starting days, around solstices and equinoxes.

1.3 The Apparent Motion of the Stars

There is an unimaginable number of stars in the universe. Yet only a tiny percentage of these astral bodies are visible, even on a clear night, with the naked eye (all naked-eye visible stars are physically similar to each other and to the Sun). The visible stars are at widely varying distances from us, but it is impossible to appreciate this fact. The brightness of the stars may be misleading in this respect: for instance Deneb and Pollux share a similar brightness but Deneb is 500 times farther away than Pollux. Distances in the universe are difficult for the human imagination to cope with—just to have an idea, sunlight takes more than eight minutes to arrive on the Earth, and the closest star beyond the Sun, Proxima Centauri, is at 4.2 light-years from us (it is not bright enough to be visible with the naked eye, though; the closest visible one is Alpha Centauri, at 4.37 light years).

To measure the brightness of a celestial object as seen from the Earth—the only brightness of concern to us here—it is customary (essentially for historical reasons) to use a rather clumsy and laborious measure called *apparent magnitude*. Unfortunately we have to make its acquaintance, since it is of universal use. The apparent magnitude is a dimensionless number m, which is such that the brighter an object appears, the lower m will be. Clearly, the Sun will thus have the *lowest* possible magnitude, which is fixed at (don't ask) −26.74. After the Sun, there are the Moon, Venus, and the brightest star, Sirius. Magnitude is measured on a logarithmic scale which has the following effect: an increase of 1 in the magnitude scale corresponds to a decrease in brightness by a factor of about 2.5. For instance, the star Vega (magnitude zero) is about 2.5 brighter than the star Antares (magnitude one).

Since we are interested only in naked eye observations, it is important to establish the limit of human visual perception. It is generally assumed that a very experienced astronomer in very favourable conditions is able to distinguish a 6 magnitude star. This is, however, to be considered as an ideal limit; normal people (including yours truly) actually have difficulty in distinguishing stars of magnitude 5, even in very favourable conditions (it goes without saying that from populated areas or towns the best we can hope to see are the first magnitude stars). In any case, in a clear sky there are typically hundreds, and sometimes thousands, of stars below magnitude 6. It follows that the sky of our ancestors was spectacularly ablaze with brilliant stars.

Today we employ a rather standardised and aseptic way of counting and naming the stars: by convention each star is identified by a progressive Greek letter plus the name (or abbreviation of the name) of the corresponding constellation, separated by a hyphen; usually, the alpha star of a constellation is also the most brilliant one. A great deal of stars, however, also have their own individual names, many of which derive from old traditions. For instance, the names of the brightest stars—in order of increasing magnitude—are Sirius (alpha-canis major), Canopus (alpha-carinae), Rigil (alpha-centauri), Arcturus (alpha-bootes), Vega (alpha-lyrae) and Capella (alpha-aurigae).

To understand the motion of the stars, we have to realise that stars move in relation to each other and also in relation to us, but that we cannot appreciate such movements in a human lifespan or even over the course of a human civilisation (at least for the periods they lasted in the past…) since even the most appreciable of such "proper" motions can be measured from the Earth in terms of a few arcminutes per millennium. So the movement of the stars as seen from the Earth is only *apparent,* due to the rotation of the Earth itself, and as such it is rigid: the relative distances between stars remain constant. To understand how the stars are perceived to move, let us first look north. If we are in a historical period during which the north celestial pole is located in an area where there is a star, this star will be seen as fixed. Then we enlarge the view concentrically from the pole. All stars which are sufficiently near the pole—so that their maximal altitude does not exceed double the pole altitude—are seen to rotate around it, never going beyond the horizon in their rotation. These stars are termed circumpolar, and are visible the whole night, every night. This immediately implies the existence of a group of stars which are sufficiently near the *opposite* pole, and which accordingly never rise. In other words, the visible portion of the heavenly vault depends on the position of the observer, and all the non-circumpolar stars visible to an observer rise and set. It follows that on certain days a certain star may only be "visible" during the hours of sunlight, or, we might say, be in conjunction with the Sun, thus effectively being invisible due to the presence of the sunlight. Stars consequently have an invisibility period (which can shrink to practically zero days, for stars very near the circumpolar ones, or be, on the contrary, extremely long, for stars rising close to due south) (Fig. 1.7).

The movement of a generic star over the course of the year is therefore characterised by a series of "heliacal phenomena". To understand the meaning of this let us follow a star starting from the day of the *heliacal rising,* i.e. the day when the conjunction with the Sun ends and the star is visible for a few moments, low on the eastern horizon, in the glow of pre-dawn. Every day, the star will rise a little earlier. The day after heliacal rising, the star will thus be visible a little longer, the next day a little longer than that, and so on, as the time of rising starts to move towards the time of sunset. On a certain day of the year, the rising is no longer visible, as the star is already seen at a certain

Fig. 1.7 The apparent motion of the stars ⊙

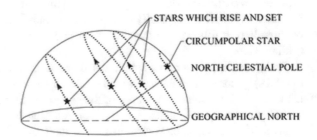

height when the Sun sets. The setting time then moves towards the hours of sunset up to the final day, when the star can only be seen to set shortly after sunset. This is called heliacal setting. The cycle of visibility/invisibility then will resume.

The day of a star's heliacal rising was of interest to many ancient cultures. Today, the phenomenon can be defined in strictly technical terms. In practice, though, it is very difficult to identify a single specific day of the year on which heliacal rising occurs, since visibility depends (apart from the local horizon, of course) on atmospheric factors, on the skills of the observer, and on the presence/absence of the Moon. It is therefore much better to think always in terms of a span of a few days. A rule of thumb which can be used to understand when an m-magnitude star is undergoing heliacal rising is to ascertain on which day the star is at its minimal visible altitude (which is greater than zero in general, see Sect. 1.10) and the Sun is (of course under the horizon) at an altitude -4 m. For instance, the bright (m = 1.5) star Castor of the constellation Gemini is (theoretically) first visible at about $1.5°$ of altitude when the Sun is about $6°$ below the horizon. The first visibility will, however, usually be a few days later than the day thus identified, and at a slightly greater altitude (so—and this matters in archaeoastronomy—also at slightly greater azimuth).

1.4   Constellations

While watching the diurnal sky, it is natural to identify forms and images in the shapes of the clouds. Similarly, during the night, images can be identified by joining the bright dots of the stars. In this way, a series of stylised figures, the *constellations,* are formed. Constellations are clearly useful for mapping out the sky and therefore for communicating related information (e.g. for orientation purposes), so that vitually all human groups developed a common pictorial image of a certain number of constellations. Our own tradition originated from the cultures of the ancient Mediterranean, notably the Near East (Mesopotamia) and Greece. From the Near East in particular we have a variety of written and iconographic sources, but the first attempt at a complete classification of the sky appears, as far as we know, in the poem *Phaenomena* by the Greek poet Aratus, who in turn refers to a work by Eudoxos, written in the fourth century BC, but which has not survived. These constellations were later codified by Ptolemy, who mentions 48 of them. During the Renaissance, many others were added, and finally, in 1922, the International Astronomical Union fixed the division of the celestial sphere into 88 constellations. All of them are referenced (besides pictorial images) using "skeletons" of segments and also, more simply, quadrangular zones of the sky, since of course modern astronomical instruments can see scores of stars in areas where the naked eye sees nothing, and such stars also are assigned to a constellation. The names and images of the original constellations codified in the Mesopotamian-Greek tradition and passed on by Ptolemy have been retained, adding others whose modern origin is sometimes quite evident, such as Telescopium.

Fig. 1.8 An early 18th century map of the constellations of the northern sky

Studying the constellations identified by a specific culture is very important, as the lore of the sky forms an important cultural memory. It usually contains a great deal of information about the imaginary world, as well as about religion, myth, and wisdom. However, identification of ancient constellations has proved to be problematic. The first problem lies in the fact that stars which we assign to different constellations can actually belong to the same constellation in another tradition, and, vice versa, stars belonging to the same constellation in our tradition can be split into different constellations by other cultures. Further, constellations identified by the same group of stars as in our tradition have often been seen in a completely different way by other cultures, so that it can be difficult to recognise them. For instance, the Egyptians saw a Bull foreleg in Ursa Major, the Mayans a turtle in Orion, and the Incas a plough in Scorpio. Finally, though rarely, we can discover close similarities in both star identification and interpretation: for instance, the Mayan constellation of the Scorpion was probably the same as that of the western tradition (Figs. 1.8 and 1.9).

Of particular importance among the constellations are those located on the ecliptic, the circle described by the Sun in the sky, which can be used to monitor the motion of the Sun. These constellations in fact form a background to the Sun as seen from the Earth, since our star shifts from one constellation to another in the

Fig. 1.9 Western Thebes. Some of the ancient Egyptian constellations as depicted on the ceiling of the tomb of Pharaoh Seti I in the Valley of the Kings ⓟ

course of the year. Of course, as for any other constellation, they may differ from culture to culture; for us, this set is traditionally made up of the twelve constellations called the Zodiac, but the Maya had thirteen zodiacal constellations (as we really should too, since also our constellation Ophiucus crosses the ecliptic).

Many constellations, and in particular the zodiacal ones, were identified in very ancient times. In my opinion, it is marvellous that scientists continue today to use images dating from such far-off antiquity and we are happy and proud to use them in this book too. However, the zodiac was developed at a time when no distinction was made between astronomy and astrology. The subject of astrology is a complex framework of arcane theories according to which the personality and personal destinies of people are influenced by their "birth sign" that is, the zodiacal constellation in which the Sun was located at the moment of their birth, and by the configurations of the Sun, Moon and the planets relative to it. Such a thing has no scientific basis whatsoever—it is a pure fantasy. Indeed, "birth signs" are continuously skewed, out of sync, from the very outset by the existence of the phenomenon called precession (see Sect. 1.6); and besides, the only physical effect exerted on us by the celestial bodies is gravity. The gravitational force, though directly proportional to the mass, decreases with the square of the distance. This means that, for example, the gravitational force exerted on a baby by—say—Mars at its minimal distance from the Earth is less than the gravitational force exerted by a 60-kg obstetrician located at 50 cm from the very same baby (*both* two forces being, of course, ridiculously small in themselves).

Having said that, we shall occasionally be dipping into *ancient* astrological lore, because in some cases (for instance, in Imperial Rome), a comprehension of it contributes considerably to understanding some key aspects of the relationship between power, architecture and astronomy (Ruggles 2015).

1.5   The Milky Way

The stars, like our Sun, are grouped into galaxies. There are billions of galaxies, which are gravitating systems—usually shaped like a disc—where stars rotate around a central nucleus. The Sun is no exception, so we all belong to a particular galaxy, the same one that all the stars we can see with the naked eye belong to as well. Because we are *inside* this galaxy, and because it is shaped like a disc, when we look at it we see a relatively narrow band of diffused luminosity traversing the sky: this is what we call *Milky Way*.

The Milky Way is not flat, but it is a sort of flattened object, so that a main (theoretical) galactic plane can be defined. This galactic plane is inclined by about 60° with respect to the ecliptic, so that the Galaxy and the Ecliptic as seen from the Earth cross two times, between Scorpio and Sagittarius and between Taurus and Gemini (the direction from the Sun to the centre of the galactic plane—where with all probability a silent, gigantic blackhole assures the stability of the whole system —passes through Sagittarius).

The luminous band of our galaxy as seen from the Earth with the naked eye has a maximal width around 12°, and it is not uniform: inside there are "branches" (points where the luminous stream divides into two) and dark regions, such as the so-called Great Rift and the Coalsack, where there are no visible stars and/or light from distant stars is blocked by the interstellar clouds. Leaving aside the brilliant stars which are seen within, or close to, the Milky Way (such Cygnus, Scorpio and Centaurus) the Milky Way itself is not very bright. However, in good weather conditions and with low light pollution (say, when stars of magnitude 4.5 are plainly visible) our galaxy becomes a prominent presence, a sort of celestial river which slowly flows, dividing the celestial sphere into two regions. Indeed, the resemblance of the Milky Way to a celestial river was noted by many cultures, for instance, the ancient Egyptians and the Incas. In particular, for the Incas the Milky Way was constituted of the same water that would later fall to earth as rain, and the dark zones in it were interpreted as images of animals forming "dark constellations" (see Sect. 6.3).

1.6   Precession

The Earth, like all rigid bodies free of constraints, has three rotational degrees of freedom which correspond to three rotational movements. These are the revolution around the axis, which completes in 24 h, and two others, precession and nutation,

Fig. 1.10 The path described by the North celestial pole in the course of a precessional cycle ⊙

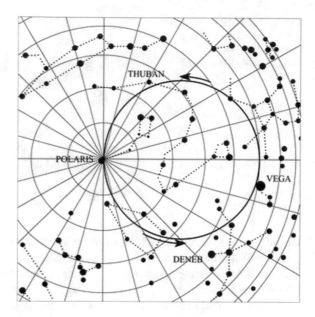

of much longer duration (to visualise them the easiest way is to think of a spinning top, with the axis circling around the vertical—precession—and describing a regular oscillation—nutation—*while* the top spins around the axis).

Of these two movements, precession—which completes a (almost closed) revolution every 25,776 years—is of fundamental importance for archaeoastronomy (Fig. 1.10).

In fact, although it has no effect on the apparent motions of the Sun, the Moon and the planets, precession changes—slowly but inexorably—the declinations of the stars. Therefore, taking precession into account is essential for a reconstruction of the nocturnal skies as ancient peoples saw them. To understand what happens, let us start from the north celestial pole. Since the Earth axis describes a circular *cone*,

Fig. 1.11 The position of the north celestial pole with respect to the stars in 2500 BC: there is a pole star, Thuban ⊙

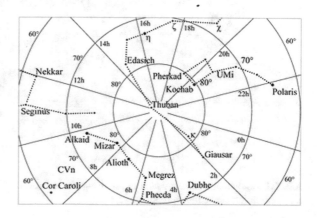

Fig. 1.12 The position of the
north celestial pole with
respect to the stars today:
there is another pole star,
Polaris

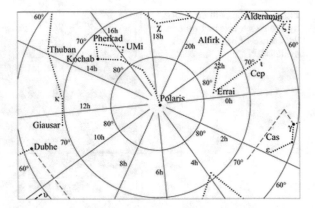

the pole moves along a circle; in a certain epoch it may or may not pass near the
position of a brilliant star. What we call the pole star today is actually the specific
star close to which the north pole is today; during the first half of the third
millennium BC another star, Thuban of the constellation Draco, was the pole star,
while in the period in between no pole star existed at all: the north celestial pole was
in a dark zone (the south pole is *always* in a dark zone, as its path never intersects a
brilliant star) (Figs. 1.11 and 1.12).

Without wishing to repeat myself ad nauseam, I should stress again that it is not the
background of stars that moves, but it is the polar axis that points to different regions
of the sky. Clearly, the declinations of all the stars are subject to this precessional shift;
in particular, some constellations can be brought under the horizon by precession
thereby proving to be, in certain epochs, constantly invisible from a certain latitude,
and eventually becoming visible again with the passing of time. One important
example is the Cross-Centaurus group, a prominent asterism of the southern

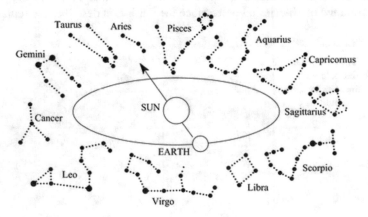

Fig. 1.13 Schematic representation of the rising sun as seen from the earth in the background of
the zodiacal constellations during the year

hemisphere, which was still visible from the Mediterranean—although low on the southern horizon—up to the last millennium BC.

As precession changes the declinations of the stars over the centuries, one of its effects is that it changes the background to the stars where the Sun rises on a fixed day of the solar year; in particular, of course, this occurs at solstices and equinoxes. For instance, in the centuries around 1AD the winter solstice was moving from Capricorn to Sagittarius (*repetita juvant*: this statement means that the winter solstice sun was starting to rise against the background of Sagittarius), and the spring equinox from Aries to Fishes. It is from the shift of the constellation housing the Spring equinox that the term "precession of the equinoxes" originates, sometimes used as a (potentially misleading) synonym for precession (Fig. 1.13).

The issue of when was precession discovered is quite delicate. Indeed, it is very likely that some specific effect was noticed by ancient cultures long before the official scientific discovery which occurred within the framework of Greek science Magli (2009). However, and in spite of worldwide "New Age" claims (the presumed New Age should be precisely the one that is triggered by the spring equinox going from Pisces to Aquarius in our own era), there is no evidence that any special relevance was attached to this discovery in ancient times. Another point which is sometimes misunderstood is that modern-day constellations are defined as regions of the sky with definite boundaries, so one can confidently state today if the Sun is rising in one constellation or another on a certain day. Equally, one could even specify precisely in which *year* the equinox moves from a constellation to another. Clearly however, this has no sense in naked eye astronomy, as the Sun can be in a region deprived of visible stars between two constellations for many decades, or even centuries. For instance, as we shall see in Sect. 10.3, Emperor Augustus chose as his personal icon the sign of Capricorn because it was culturally associated with the winter solstice as a sign of renewal, although—as mentioned above—the midwinter Sun was at those times already rising in the region between the visible stars of Capricorn and those of Sagittarius.

1.7 The Apparent Motion of the Moon

The Moon moves around the Earth on an elliptical orbit. The plane containing the Earth and the Moon orbit is not the ecliptic, but forms with it an angle—usually denoted by i—of $5°9'$; the two points of intersections of the orbit of the Moon with the Ecliptic are called nodes. The period of revolution of the Moon is equal to its period of rotation so that an observer on the Earth can see only one face of the Moon (actually slightly more than one half). So, incredible as it may seem, the first time that humans were able to view the other side of the Moon was as recent as 1959, with the photographs taken by the Russian probe Luna 3.

Of course, one half of the lunar surface is always illuminated by the Sun. This means that, of the face of the Moon "reserved" for us, we can see on each day only

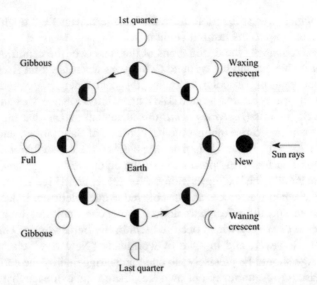

Fig. 1.14 A schematic, two-dimensional representation of the phases of the Moon. The phases are generated by the different illumination of the half of the Moon which can be seen from the earth (since in reality the Moon, the earth and the Sun do not belong to the same plane, we do not experience eclipses every lunar month, as instead the image could suggest)

that part belonging to that half which is illuminated by the Sun. This provokes the lunar phases, with a cycle of approximately 29.5 days (Fig. 1.14).

The sidereal month of the Moon lasts, on the contrary, 27.2 days, in the course of which the rising point of the Moon on the horizon follows a cycle similar to that made by the Sun over the course of a solar year, that is, with one rising extreme located north of east and the other, symmetrical, at the south of east. So we can talk about northern and southern lunar "standstills", in analogy with the two solstices. Unlike with what happens to the Sun, however, (for which the declinations at the extrema are always equal to $\pm\varepsilon$) the declinations of the Moon at standstills vary. As a result of various physical factors, in fact, the line of the nodes revolves clockwise in relation to the Moon, completing a cycle every 18.61 years. Thanks to this phenomenon, the declination of the Moon at the northern lunar standstill oscillates from a maximum equal to $\varepsilon + i$ (today thus equal to 28°39′), to a minimum equal to $\varepsilon - i$ (today 18°21′). Analogously, the declination of the Moon at the southern standstill oscillates between $-(\varepsilon + i)$ and $-(\varepsilon - i)$. Accordingly, we can speak about north and south *major and minor* lunar standstills or *lunistices* (a small effect called *wobbling* causes a further variation in declination with an amplitude of nine arc minutes and a period of 173 days). The last major standstills were reached in 2006 and therefore will be reached again in 2024. The azimuths of the Moon at the standstills of course depend on the latitude; however, it is always the case that the rising azimuth at the major northern standstill is to the north of that of the Sun at the summer solstice, and that of the minor northern standstill is to the south of it (a

Fig. 1.15 The rising-setting azimuths of the moon at a certain latitude, with the solar arc also highlighted. Independently from the latitude, the minimal lunar amplitude is always smaller than the solar arc, and corresponds to the minor lunar standstills (m). The maximal lunar amplitude is always greater than the solar arc, and corresponds to the major lunar standstills (M)

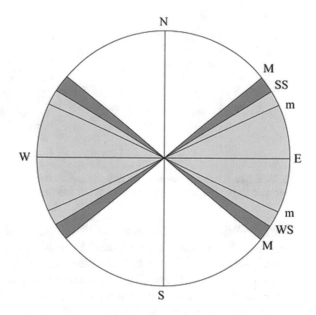

symmetrical situation will occur for the southern lunar standstills and the winter solstice) (Fig. 1.15).

The long-term variation of 18.61 years in the Moon's rising points overlaps with the cycle of phases in a complicated way. This means that a standstill can even occur at a new Moon and thus be unobservable. However, in the year of the lunistice, the Moon will have monthly maximal and minimal declinations which—although these correspond exactly to those of the lunistices once only—are all very close to them. Now, due to the characteristics of the motion of the Moon, an interesting phenomenon occurs. In fact, it always happens that the full Moon closest in time to the summer solstice is at the minimal declination, and similarly, the full Moon closest in time to the winter solstice is at maximal declination. Both spectacles are really worth seeing in the lunistice year. The full Moon near the winter solstice—the longest night of the year—culminates very high in the sky (higher than the Sun at summer solstice, in particular) and remains in the sky the longest; it can even become circumpolar at high latitudes. The full Moon near the summer solstice—the shortest night of the year—culminates at its lowest altitude ever. When we consider the possibility that ancient astronomers and architects might have been interested in the extreme positions of the Moon, it looks advisable to think of a kind of general interest in these (fully visible and spectacular) phenomena, rather than of extremely precise measurements of lunar azimuths at standstills and/or to different phases (see however Sims 2011) (Fig. 1.16).

Other spectacular events associated with the Moon are of course the *eclipses*. In a solar eclipse, the Moon passes between the Sun and the Earth. A solar eclipse can thus occur only if the Moon passes a node around the time of new Moon, since it is the illuminated face of the Moon that forms the shadow on the Earth. In a lunar

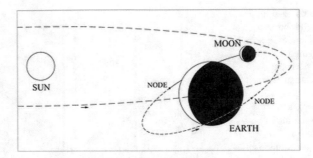

Fig. 1.16 The relative position of the Ecliptic and of the Moon orbital plane. Eclipses can only occur near the nodes, namely the intersections of the two planes

eclipse the Earth casts a shadow on the illuminated face of the Moon, so it can only occur if the Moon passes a node around the time of a full Moon. Solar eclipses are generally witnessed only as a portion of the Sun's disk being covered by the Moon, or partial eclipses. However, by a curious coincidence, the apparent sizes of the Moon and of the Sun from the Earth are roughly equal, and therefore there usually exists a relatively narrow (not more than 250 km wide) strip of the Earth's surface (called the path of totality) where the Moon projects its full shadow and the Sun is totally eclipsed (sometimes the Moon's apparent size—due to the slight variation of its distance from the Earth—is too small to cover the Sun completely and a so-called annular eclipse takes place). Unlike solar eclipses, lunar eclipses are long-lasting and easy-to-view phenomena.

Eclipses are irrelevant with regard to the alignments of buildings; however, there is no doubt that many ancient cultures had an interest in them: they were accurately recorded (for instance in Chinese annals) and the Mayans devised methods to predict on which days eclipses were likely to occur (see Aveni (2001) for a complete discussion).

1.8   The Apparent Motion of the Planets

As far as archaeoastronomy is concerned, the only planet for which any kind of connection with astronomically oriented buildings has been reasonably argued is Venus. However, for completeness, a brief overview of the motion of all the visible planets will now be given.

The solar system is a complex collection of objects, none of which emits its own light, all gravitating around the Sun. Some have a relatively simple, elliptic orbit (I will say, with a measure of astronomer's licence, that such objects move around the Sun) while others have a more complicated orbit, such as the Moon, which moves around the Earth while the Earth moves around the Sun. The planets are the greatest of the "simple-moving" objects; in order of distance from the Sun they are:

Mercury, Venus, Earth, Mars, Jupiter, Saturn. The remaining planets Uranus, Neptune and Pluto are not visible with the naked eye (Uranus is on the threshold of visibility at $m = 6$). The planes containing the orbit of each planet (and thus the planets and the Sun) are inclined with respect to the ecliptic, but the inclinations are small; the greatest are those of Mercury ($7°$) and Venus ($3°24'$).

The planets themselves are vastly different from each other. For instance, Mercury has a very eccentric orbit, is as dense as the Earth, and is otherwise similar to the Moon, while Venus displays little eccentricity, has a composition similar to that of the Earth, but also a formidable atmosphere of dense carbon dioxide, with clouds of sulphuric acid. But for our purposes the main point here is to note the distinction between inner and outer planets. Indeed, since we are on the Earth, the motions of the inner planets Mercury and Venus, and that of the outer planets, Mars, Jupiter and Saturn, share special features.

Let us begin with Venus. The time it takes Venus to make a complete revolution on its orbit around the Sun (sidereal period) is 225 days, but clearly the Earth also revolves around the Sun, so from our viewpoint the planet actually reappears in the same configuration in relation to the Sun after a different period, called synodic. This period is approximately 584 days (583.92 days). During each cycle, Venus will be invisible every time the Sun's light is in conjunction with it; one would be tempted to say, every time it is behind or in front of the Sun, although of course this does not mean that they lie precisely on the same plane (sometimes Venus really does eclipse the Sun as a black dot passing over the Sun's surface). The periods of disappearance from sight are about 50 and 8 days, respectively. The two periods of visibility correspond to the planet appearing as the *Evening Star*, located in the west immediately after sunset, and as the *Morning Star,* located in the east before sunrise, for about 268 and 258 days respectively (Fig. 1.17).

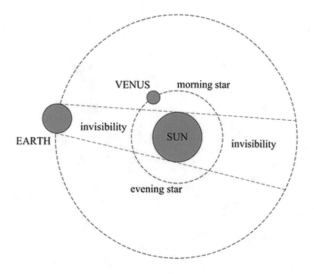

Fig. 1.17 Schematic, bi-dimensional representation of the cycle of visibility/ invisibility of Venus

In the course of this cycle Venus as seen from the Earth is subject to a change of declination and, therefore, of azimuth at rising. The inclination v = 3°24' of the plane of the Venus orbit has the effect of enabling the planet to attain a maximal/minimal declination of ±(ε + v), that is, ±27° today, increasing a little (due to the variation of ε) in the past. To such declinations correspond azimuths which lie between the azimuths of the Sun at the summer/winter solstice and those of the Moon at the major northern/southern standstill. Due to various physical effects, however, Venus attains the maximal possible declination only once every five synodic cycles; since 584 × 5 = 365 × 8, it can be said that the Venus extreme declinations have a cycle of 8 years. During the last four millennia, both the maximum and the minimum declinations have been attained during the planet's evening visibility (so that the corresponding azimuths were observable at setting), and have been seasonally related, in the sense that they usually occurred in late April/early May, a period which is of the utmost importance the agriculture in

Fig. 1.18 Schematic explanation of the retrograde motion of a exterior planet. The lines represents the line of sight from the earth to the planet in subsequent, few days; they "hit" the fixed stars background in the corresponding numbered points

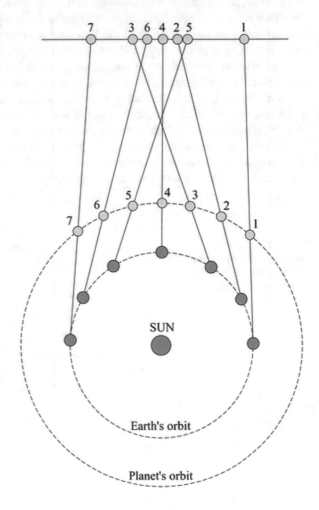

Mesoamerica, marking as it does the onset of the rainy season. There are texts which unquestionably show the interest of Mesomaerican cultures in the cycle of Venus, and also convincing cases of orientations to this planet, as we shall see in Sect. 9.1. As far as Mercury is concerned, its behaviour is qualitatively similar, but the planet can only be seen for shorter periods of time Kelley and Milone (2005).

The alternating behaviour of Venus as the Morning/Evening star, shared also by Mercury, is certainly the main characteristic of its apparent motion. The most evident behaviour which characterises the motion of the outer planets visible to the naked eye, Mars, Jupiter and Saturn, is another one. Consider, to have a clearer idea of the concepts in play, Mars. This planet also has a period of conjunction with the Sun, and therefore of invisibility (about 120 days). After invisibility, it is subject to heliacal rising and rises progressively earlier, like a star. At a certain point, the Earth passes between the planet and the Sun, that is, our planet runs a section of its orbit that lies between the Sun and the planet's orbit. This is the period in which Mars is both closest and brightest. The Earth then eventually overtakes Mars. Both of course continue to proceed forward on their orbit, but what the observer on the Earth will actually see is a little odd. In fact, like any celestial object, Mars continues to regularly travel from east to west each night. However, if we project the sight line of the Earth's observer beyond Mars, we see that the star's background "turns back". If, therefore, the Earth-based observer keeps a note of the background of the stars each night, Mars will be seen to "move back" for a certain number of days, until the overtaking is complete. This period lasts about 75 days and is the so-called *retrograde motion* (Fig. 1.18).

The ideal centre of retrograde motion occurs when Mars is opposite the Sun; if the position of the planet is plotted as a function of the days, the corresponding curve is a loop in the sky covering about one half of a zodiacal constellation.

As far as the rising/setting positions of planets apart from Venus are concerned, as mentioned, we have no architectural evidence of any interest on the part of ancient cultures; indeed, it would be difficult to have any such proof since the maximal declinations of these planets are reached irregularly and are too close to that of the Sun to imply relevant distinctions in azimuths, and therefore a hypothetical alignment towards—say—the maximal declination of Jupiter would be virtually indistinguishable from a summer solstice alignment to the Sun. However, we do have quite convincing written evidence of a connection of the planets with power and religion in many ancient cultures, from the Babylonians to the Mayas.

1.9 Calendars

As we shall soon see, the control of the flow of time through the establishment of, and constant attention to, a calendar was not only fundamental for practical (e.g. agricultural) reasons, but was deeply and intimately connected with the exercise of power in many ancient cultures. As a consequence, the history of the anthropological and cognitive aspects of calendrics is a delicate topic which, in my view, still awaits a comprehensive, global study. Certainly this is not the place for this, but it is impossible to learn about archaeoastronomy without having a grasp of the basics of the history of calendars, and especially of that of *our* calendar, the Gregorian, because, naturally, we need a common language to discuss dates in various places and epochs. So, let us first seek to understand how the Gregorian works.

The time that the Sun takes to return to the same position, as seen from Earth, is called a tropical year. It seems obvious for us to measure it in terms of days, but it may be equally obvious that there is no reason whatsoever it should last an *integer* number of days and indeed, it is not so: the tropical year lasts 365.2421897 days. *If* we want to count time with a calendar tuned to the cycle of the Sun, we need an approximation. This leads to the *Julian* calendar, introduced in Rome by Julius Caesar: a calendar of 365 days per year in which each four years a "leap" year of 366 days is added (giving an average of 365.25–365 days 6 h—for the tropical year). Due to the approximation made, the Julian calendar drifts with respect to the solar cycle by about 10 min per year, that is, about one day each 144 years, so it took many centuries for the difference to become macroscopic. The Julian was replaced by the current Gregorian calendar by Pope Gregory XIII, who introduced it in 1582. This calendar, which corresponds to an average length of the year of 365.2425 days (365 days 5 h 49 min 12 s), is practically perfect over the span of several millennia, so we can safely speak of Gregorian dates as coinciding with tropical dates throughout the whole of human civilisation (note, however, that most astronomical software uses the Julian calendar for dates before the Gregorian reform; moreover, they use a "proleptic" Julian calendar—that is, reversed—for dates before the introduction of the Julian). The way the adjustment of the Gregorian years is effected is very ingenuous: the Julian leap years are retained, but reducing the number of them in four centuries from 100 to 97, thereby gaining 3 days *less* each 400 years. The rule by which it is decided which leap-years are to be skipped is: centurial years are leap years only if they are exactly divisible by 400 (so that the years 1700, 1800, and 1900 were not leap years, whereas the year 2000 was).

The Julian and Gregorian calendars were devised by astronomers who wished to synchronise the count of the days of the year with the tropical year and hence with the seasons. Yet theoretically one can schedule and manage all the important activities associated with seasons—e.g. harvesting and sowing the fields—by means of other instruments, for instance, using astronomical alignments which point to the rising or setting of the Sun, or using other celestial phenomena, such as the Heliacal rising of bright stars. This is not always clearly explained in works on calendars, since our conception of calendar as a count which is strictly underpinned by the physical reality of the solar cycle is taken for granted as the most "evolved". It is, therefore, important to stress that there is *absolutely nothing* that obliges a human civilisation to have a solar-tuned calendar, nor is there any reason why a solar-tuned calendar should be considered more evolved or sophisticated than any other. As a matter of fact, many great civilisations of the past—including the Egyptians and the Maya—opted for a calendar consisting of a fixed number of days, 365. Such a calendar very rapidly wanders out of sync with the seasons, taking some 1460 years to realign, but this was unimportant for them, as we have ample proof that both these civilisations had a very clear conception of the fractional length in days of the tropical year as well as efficient ways of keeping the shift of the calendar under control when necessary for practical applications.

Besides the Sun, other recursive celestial phenomena are suitable for keeping track of the passing of time; in particular, those connected with the Moon. A lunar calendar is most naturally based on the phase cycle, although a calendar based on sidereal months may also be devised. In a phase-based calendar the fundamental cycle is a lunation, which naturally defines a lunar month. A purely lunar calendar might of course have a year comprising an arbitrary number of such months. However, most lunar calendars devised in antiquity are actually luni-solar that is, add a mechanism of correction to re-align them with the solar year. A way of doing this is to observe that 12 lunations correspond to 354.37 days, so that a delay of about ten days will be accumulated each solar year. Therefore, many luni-solar calendars have 12 lunar months but add a thirteenth months each 3 years; this occurs, for instance, in the Chinese and the Hindu calendars (the Islamic calendar, although also based on 12 months, is purely lunar). The beginning of the month of a lunar calendar can be identified in different ways; for instance, the first day of a month might be the day after the full Moon or that of the new Moon itself, or also that of the first sighting of a lunar crescent.

1.10   The Observation of Celestial Bodies on the Horizon

When speaking about naked-eye astronomical observations we have always assumed "perfect" conditions of visibility. The Earth, however, is surrounded by the atmosphere: a mixture of gases (mainly nitrogen 78% and oxygen 21% but also water vapour, 1% on average). About 80% of this gaseous mass is concentrated in the first layer, the troposphere, which is only 12 km thick. The atmosphere acts as an optical lens, whose physical properties are not constant as they depend on temperature and pressure. Furthermore, the physical characteristics of the atmosphere were drastically altered (and are still being altered) by humanity by the spread of large-scale pollution following the industrial revolution.

The atmospheric effects on the light coming from a celestial source depend strongly on how much of the atmospheric lens the light rays have to travel within. Since the atmosphere is very thin, it is clear that these effects will become smaller the higher the celestial object whose light we are observing is, becoming fully negligible at, say, 20° of altitude. However, in archaeoastronomy we are interested mainly in horizon phenomena, so the section of atmosphere the light has to travel in is the maximal one. This leads to two important effects, *extinction* and *refraction*.

Extinction is the absorption and scattering of electromagnetic radiation—that is, light—by the atmospheric gas. It affects the color of the source—which becomes redder—and the perceived brightness, the apparent magnitude. Extinction is measured by an extinction coefficient k which varies in accordance with many factors such as the altitude (being higher at sea level) the latitude, the temperature and so on. High values of k—in the range 0.20–0.25—are observed in polluted areas and lead to very strong attenuation in magnitude for objects close to the horizon; it is, however, reasonable to suppose that (average) conditions were superior in the past (in particular, before the industrial revolution). Therefore, in archaeoastronomy, it would seem acceptable to use a lower range (from 0.14 in optimal conditions, to 0.17), which can today be attained only in very special observation places. Having said this, we need a way of taking extinction into account without, of course, having any way of knowing what real atmospheric conditions were like when the alignments we are studying were devised. To this end, Alexander Thom proposed the following rule of thumb: the altitude in degrees at which an object can be first be seen with the naked eye—that is, the altitude at which his magnitude corrected by extinction is less than 6—is equal to the apparent magnitude expressed in degrees. For instance, a 1 magnitude star is visible at 1°, a 4 magnitude star (or group of stars having an overall magnitude equivalent to this, such as, for instance, the Pleiades) would be visible at an altitude not less than 4°. Thom's rule must be taken as a optimistic limit—that is, a star will certainly *not* be visible at a lower altitude if we refer to this rule Schaefer (1986).

A second effect caused by the atmosphere is the deviation of light rays, called refraction. Refraction is the phenomenon experienced, for instance, when one sees the apparent bending of a oar in water. Atmospheric refraction tends to lift the image of the source above the horizon, so that celestial objects appear higher in the sky than they are in reality. There is, therefore, an additional refraction correction to the altitudes, which also affects the values of the azimuth. Refraction, like extinction, depends on local atmospheric conditions and on latitude, and a precise estimate is not easy to make. In any case, it is again negligible at high altitude, reaching some 10′ at 5° and a maximum at the horizon, where the correction is of +34′. Incidentally, since 34′ is greater than the diameter of the Sun, when we see our star rising, the unrefracted Sun is still actually below the horizon.

A final effect which has to be taken into account for observation at the horizon is parallax, which is caused by the displaced position of the observer on the Earth surface in relation to the direct line to the celestial object. For the purposes of naked-eye observations, parallax correction is (very) important for the Moon, because of its proximity to the Earth. The parallax correction to the altitude of the Moon at the horizon is 57′.

References

Aveni, A. F. (2001). *Skywatchers: A revised and updated version of Skywatchers of ancient Mexico*. Austin: University of Texas Press.

Kelley, D., & Milone, E. (2005). *Exploring ancient skies: An encyclopedic survey of archaeoastronomy*. NY: Springer.

Magli, G. (2009). *Mysteries and discoveries of archaeoastronomy*. NY: Springer-Verlag.

Ruggles, C. (Ed.). (2015). *Handbook of archaeoastronomy and ethnoastronomy*. NY: Springer-Verlag.

Schaefer, B. (1986). Atmospheric extinction effects on stellar alignments. *Archaeoastronomy, 10,* 32–42.

Sims, L. (2011). Where is cultural astronomy going? In F. Pimenta, N. Ribeiro, F. Silva, N. Campion, A. Joaquinito, & L. Tirapicos (Eds.), *SEAC 2011 Stars and Stones: Voyages in Archaeoastronomy and Cultural Astronomy.* London: BAR.

Chapter 2
Acquiring Data

2.1 Archaeoastronomy Fieldwork

Ancient peoples' attitudes towards, and relationships with, the natural environment
and the landscape were often completely at odds from ours. Two fundamental concepts must be borne in mind, in particular. The first is that ancient man was a religious
man, and the second is that religion was bound up with the natural cycles, and these
natural cycles were bound up with power. We shall gradually gain more insight into
these fundamental concepts as this book progresses, but it is essential to take account
of them when engaged in fieldwork because they were *reflected in the landscapes.*
Typically the monuments which are the object of study of archaeoastronomy are
immersed in landscapes which were of special, sacred significance for the people who
built them. Therefore, the first step in any kind of archaeoastronomy fieldwork is to
look around carefully, trying to understand how the ancient configuration of the
surroundings might have been. This might be relatively easy if the environment has
remained unspoilt or, on the contrary, depressingly difficult if humans have modified
the landscape beyond recognition. Furthermore, we must try to imagine the landscape
and the sky as a whole, since it is in this way that ancient people perceived the Cosmos
around them. This may also be problematic, due to environmental pollution, light
pollution, and—last but not least—precession. Fortunately, as we shall see in the next
chapter, computers may be of enormous assistance in tackling these tasks through the
use of virtual globes and planetarium software.

The second main point, after seeking to appreciate each site as a whole, is to
make an accurate relief of the visible horizon. This has an obvious technical purpose, namely, to allow the calculation of the declinations relating to directions of
interest through alt-azimuthal coordinates. In this respect, the role of this relief is
fundamental: it may happen, for example, that an azimuth which is many degrees to
the south of the azimuth of the rising Sun at the winter solstice with a flat horizon
may correspond precisely to the solstice if the horizon itself is occupied by
mountains high enough to hide the climbing Sun behind them, up to when our star

© Springer Nature Switzerland AG 2020
G. Magli, *Archaeoastronomy*, Undergraduate Lecture Notes in Physics,
https://doi.org/10.1007/978-3-030-45147-9_2

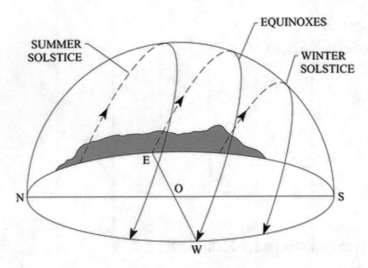

Fig. 2.1 The effect of a non flat horizon on the local observations of the rising sun ▶

has an altitude—and therefore an azimuth—much greater than those at rising (see the example of Aosta in Sect. 10.3) (Fig. 2.1).

The relief of the horizon is also important for another reason, since prominent features may be of cultural/sociological interest, such as sacred mountains, or special profiles which were considered sacred in ancient times. Additionally, the horizon may be occupied by other monuments that are culturally interrelated, and profiles may display typical breaks (notches or peaks) that might have been used as distant foresights to observe astronomical phenomena.

The third main point is to pay attention to as much detail as possible. Bring an old-fashioned paper notebook, and make sketches of anything that grabs your attention, taking notes as you go along. Bring a digital camera and take pictures from all angles; more generally, try to record all things which might turn out to be significant in the follow-up analysis. Even armed with all this, however, further visits will inevitably be necessary after the first data analysis of a site has been done.

Finally, you are ready for the relief of the possible alignments using the techniques described below. For this relief, whatever instrument you use, the first step is to find north. The second step is to be *sure* you have found north. The third step is to be really, really sure that the direction you have found and you are currently calling "north" is the true north (of course, within the expected range of errors). Now, take as many measurement as possible, repeating them with different means whenever possible. For instance, always have a look at the Google Earth ruler on your PC, even if you are making a high precision on-field relief with a theodolite. Be careful about the significance you attach to a site in terms of alignment. Indeed, if you have a well-preserved Greek or Egyptian temple, then it will be easy to determine the main axis of the temple and this alignment will obviously be the principal one to be measured. But if the monument is in a poor condition, it may not be so easy. Also, in restricted areas it is better to identify ranges rather than single alignments; for instance, if you are measuring a row of a few megaliths, it is not easy—or it may even be impossible—to define a single line of sight which "crosses the centre" of all the stones (Ruggles 1999).

It might be in order here to explain few notions relating to the acquisition of experimental data in physics. It should be said that nothing can be measured with *absolute* precision. First of all, any instrument has an intrinsic limit, beyond which we cannot measure. For instance, if we measure the side of a table with a ruler graduated in centimetres, we can only claim to know its length within ± 1 cm. In any measure, one defines *accuracy* as the degree of proximity of the result to the true value of the sought quantity. Accuracy is different from *precision* which is the degree to which repeated measures in the same experimental conditions give close values. A diligent experimenter must do his best to improve both by eliminating sources of error. To understand this, we can use a simple analogy. If I am a good archer, my arrows will all reach the target in a restricted zone (high precision). But if my bow is not perfectly balanced, this zone will not be in the centre of the target (low accuracy). Correspondingly, there are two different sources of errors operating here. One is systematic: it is the fact that the bow is not balanced. The other is random error: on the basis of factors which vary slightly with each bow, even the most skilful archer will not send all the arrows exactly to the same point. Balancing the bow will thus be in order, and making several shots will identify better the area of the target. Analogously, in archaeoastronomy it is always advisable to repeat measurements several times, and to attempt to eliminate all systematic errors.

2.2 The Magnetic Compass

The Earth possesses a magnetic field which effectively makes our planet a huge magnet. The poles of this magnet correspond broadly to the Earth poles, and the force lines of the magnetic field roughly correspond to the meridians. Since the magnetic field is neither constant nor uniform, an iron needle rotating freely on the Earth's surface aligns in a direction—called *magnetic north*—which depends on the position of the observer as well as on time. The deviation of the direction of magnetic north with respect to the geographical north, in a certain place and at the certain time, is called magnetic declination (Fig. 2.2).

A (magnetic) compass is an instrument based on a magnetised pointer turning freely upon a pivot, in air or in a stabilising fluid. When the compass is held level, the needle turns until it stabilises, pointing toward magnetic north. The compass is equipped with a circular grade scale which allows for the determination of azimuths in relation to the magnetic north. For archaeoastronomy, a *bearing* compass—that is, a magnetic compass mounted in such a way that it allows taking the bearings of objects by sighting them with the lubber line interposed—is worth using.

The compass is a very simple, relatively cheap and pocket-sized instrument, which works without any specific preparation, under any atmospheric conditions, and as such it should be an inseparable companion for any archaeoastronomer.

Fig. 2.2 The magnetic
declination

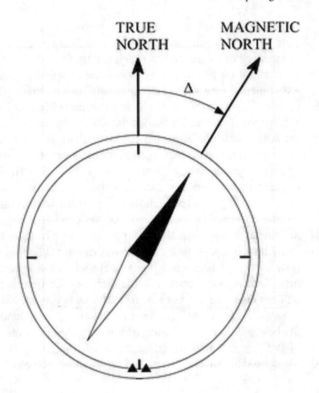

It allows us, in particular, to have a first snapshot of the orientations on a site. What
is more, if used with due caution, it can also be employed for scientific surveys.
Caution should be taken with the following:

- as with any instrument of measurement, your compass might have a
  non-negligible intrinsic error (nothing to do with the error due to magnetic
  declination, which is addressed below). Check for the presence of an intrinsic
  error by making a double-blind measure of a trial azimuth with a high accuracy
  method (such as a theodolite, or using an already well-measured alignment).
- as with any measurement, always repeat the bearings several times and take
  averages. For instance, if you want to measure the axis of a temple, measure
  from both extremes; if you are working with a colleague, take four measurements, and so on.
- the compass is sensitive to iron. Iron can be naturally present in the soil and may
  cause magnetic anomalies, and consequently, wrong readings. Also beware of:
  iron bars in fencing around archaeological sites, metal bars in restoration
  scaffolding, iron in the archaeoastronomer's wristwatch or glasses etc. The
  compass is also sensitive to electromagnetic fields of any nature (which may be
  generated by electric wires or personal computers). Information about the
  possible presence of local magnetic anomalies should be obtained, and the
  utmost care taken to avoid any distorting influences.

Once a set of compass measurements has been carefully obtained, the database must be accurately adjusted for the systematic error due to magnetic declination. Sometimes the value of magnetic declination appears on official maps, together with an estimate of future variations. This kind of information is, however, seldom reliable and in any case *insufficiently accurate* for archaeoastronomy studies. Just do not use it. Fortunately, the National Geophysical Data Center of the National Oceanic and Atmospheric Administration of the United States provides a very efficient free online calculator, which—on the basis of a mathematical model of the Earth's magnetic field—provides the magnetic declination once geographical coordinates and time have been inserted. Suppose, for example, that the value given for the time the measurements were taken is 4 degrees east. This means that any reading you took indicated an azimuth with a "zero" point 4° greater than the true zero; accordingly, you must add 4° to any measure. If the declination is, say, 5° west, the bearing exceeds the true one so, in other words, you must subtract 5° from any measure.

2.3 The Clinometer

The second set of data required in archaeoastronomy fieldwork is the altitude of the visible horizon in relation to all the azimuths of interest. In archaeoastronomy we are not so concerned with the distance of objects, but rather, we need to know to what extent they occupy the view. In other words, we are interested in angular heights: a hill which is 100 m high but is very near can block the view much more than a distant lofty mountain. To measure such *apparent* heights one needs to measure angles of sights. A clinometer is a simple instrument for measuring such angles, which works—like the compass—under any climatic conditions, and as such is quite adequate for a preliminary and/or quick survey of a site (sometimes compass and clinometer are combined in a tandem instrument) (Fig. 2.3).

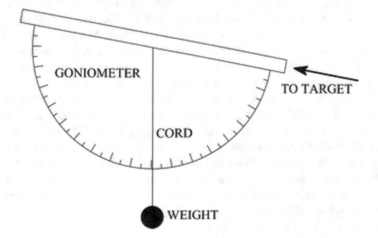

Fig. 2.3 A clinometer is essentially a goniometer used vertically, as shown in this picture ⊙

Essentially, a clinometer is a goniometer used vertically, and indeed a simple clinometer can be constructed by using a half-goniometer (a graduated semicircle). Fix the base of the semicircle on a stick and let a plumb cord hang freely from the centre of the circumference. Then place the stick in line with the eye and point at (or to be more exact, look through) the edge of the object you want to measure. Then the cord will "cross" the graduated semicircle at a certain value. The apparent altitude in degrees of the sighted object is obtained by subtracting this value from 90°. Professional clinometers employ the same principle, but work with a disc which is able to rotate freely in an oil bath. Standard professional clinometers (such as those which can be purchased combined with professional compasses, like precision compass-clinometer instruments), if used carefully, can attain a precision of ¼°. It goes without saying that it is better to repeat several times also height measurements, recalling however, of course, that they depend on the position from which they were taken. For instance, while you can measure the azimuth of a straight line (like the axis of a temple) from any point along it, the horizon height will be different from point to point.

2.4   The Theodolite

One may often chance to see on roads or in construction sites (and also in archaeological excavations) two people who are working, one with an optical instrument mounted on a tripod, the other with a marked post equipped with a reflection signal. They are performing a survey using a *theodolite*, an optical instrument for measuring angles in horizontal and vertical planes.

A theodolite is essentially a telescope which is turnable both horizontally and vertically, fitted with scales which allow the measurement of both angles relative to the sighted point with a excellent precision, typically better than one arc minute. In today's theodolites, readings are usually electronic, and an integrated electro-optical distance measuring device, generally infrared-based, allows the measurement of distances and thus a 3D mapping. These integrated instruments are called total stations. Data can be registered electronically and downloaded in external devices.

Clearly a theodolite is the most suitable instrument for archaeoastronomical measurements (azimuths and horizon heights) (Aveni 2001). However, it can provide very reliable results only if the data, which a priori refer to the instrument's arbitrary reference system of coordinates, can be related to geographical data, that is —yes—if true north has been ascertained accurately. So, without wishing to state the obvious, it should be reiterated that using a very precise instrument like a total station is useless in archaeoastronomy if the geographical north has not been determined with comparable precision. Since, however, an approximate north usually suffices for standard theodolite uses in civil engineering, it may be the case that even trained theodolite surveyors are not acquainted with the methods used for a precise determination of true north. So I shall briefly describe how to proceed with such a determination (Aveni 2001).

The idea is based on the fact that it is possible to discover with great accuracy the position of the Sun in the sky at any time with the tables called ephemeris, where the Sun's azimuth and altitude are tabulated at any place for any given date and time, or with appropriate computer software. Therefore, by knowing at what instant of time a measurement of the Sun is taken, we immediately know at which azimuth in relation to true north the reading corresponds and, accordingly, set the zero of the instrument. It follows that the theodolite can effectively be used only on days in which at least a good twenty/thirty of minutes of sunshine, most preferably close to noon, are available on the site. Moreover, a radio-chronometer will be needed. In practice, one first determines an arbitrary zero-scale by measuring the azimuth of a fixed, possibly distant and elevated object (antenna, electric wire, bell tower or the like). All measurements will be taken, with the high precision of which the instrument is usually capable, in relation to this *arbitrary* zero. Once the true azimuth of the reference object is known with respect to north, this value will be used to correct all other measures. To watch the Sun, the instrument must be equipped with a specifically guaranteed sun filter on the lens; looking directly at the Sun any other way is potentially harmful both for the instrument and for the surveyor's eyes. It is always advisable to perform this procedure during the central hours of the day. Indeed, because of refraction, measurements in the first/last hours of the day may be subject to errors. In this respect, it is important to remember that time is conventionally the same on each time zone, but the true time at which the Sun culminates—the local noon—varies within the time zone and depends on the specific latitude. It is, therefore, essential to have a precise determination of the geographical coordinates of the site, which can be obtained through GPS reading (see below). Once everything is ready, the operator centres the Sun in the theodolite viewing grid while watching the chronometer or, even better, while an assistant calls out the time from the chronometer. Successive readings can now be taken and averaged to obtain the azimuth with maximum accuracy.

2.5 The Global Positioning System

The Global Positioning System (GPS) is a free-access navigation system based on a cluster of satellites orbiting the Earth and completing one revolution in 12 h. The satellites (currently 32) are distributed on six different orbital planes. The system has been specifically devised in such a way that on any point of the planet's surface it is, in principle, possible to receive the signals from a number of satellites varying between five and eight. Each satellite transmits continuously two sets of data: time and satellite position at that time. To access the system one needs a GPS receiver. The receiver computes the distance to each satellite. Then, using these distances and a relatively simple algorithm, its software is able to identify the receiver's own position on Earth: latitude, longitude as well as altitude (navigation systems, such as those used in cars, also apply other algorithms, which allow the determination of other parameters like receiver speed and direction of motion). Obviously, the greater the number of satellites the receiver can track, the more accurate the measurement will be.

Several phenomena affect the precise synchronisation between each satellite and the receiver, producing errors in positioning. Of course, in archaeoastronomy we need the greatest possible accuracy, and there are techniques—based on data analysis and/or the use of differential systems correcting the incoming signals by means of data from reference receivers—which enhance the accuracy of standard GPS positioning. Another way to improve accuracy is to use an instrument able to connect both to GPS and to the Russian cluster of satellites GLONASS (which works much in the same way as GPS). The standard error of a normal GPS is a few metres, which can be reduced by up to few centimetres by proper data processing. It should be said that—even when a single economical GPS is used—it is always expedient to try to work with as many satellites as possible (traceability of satellites depends, for instance, on the presence of woods), and to let the instrument, held in a fixed position, acquire data for several minutes.

In considering the application of GPS in archaeoastronomy fieldwork, and thus the measuring of azimuths with the aid of a GPS, two different situations must be distinguished. First of all, consider the case of long distance measures—in the order, say, of several hundreds of metres or more, as might be, for example the straight roads of a Roman town, or a long ceremonial road, such as the "Street of the Dead" in Teotihuacan, Mexico. In such cases, a direct measurement with a GPS is certainly reliable: the instrument can be used to obtain the geographical coordinates of the two extrema of the alignment considered, and preferably also a few readings along the same line. Then simple trigonometry formulas (the so-called Puiseux-Weingarten transformation) can be applied to find the azimuth (Muller 2012).

Obviously, the same procedure can, in theory, also be used for buildings: for example, to measure the axis of a Greek temple. However, the error in measurement increases the shorter the distance is between the extrema, so in such cases the GPS is best applied together with another device. Again we should remember that the main point of any measurement campaign is firstly to find geographic north. To this aim GPS can be used to find the azimuth of a fixed direction (and consequently, north) using the technique described above for long alignments. It is often sufficient to identify a prominent far-off point or landmark—but one that is visible and accessible from the site in question, for example, a water reservoir, or a bell tower —and to GPS-measure this alignment first (the optimal target should be at a distance of several kilometres). The azimuth thus obtained can then be used to calibrate locally another instrument, a theodolite or even a magnetic compass.

To conclude, a few words about the choice of the most suitable instrument are in order. If used correctly, a good compass can give azimuths within ½°, while a theodolite or a differential GPS can give results within 1' (or even less, but efforts to reduce errors further would be frankly exaggerate, since a direction originally determined by the naked human eye—and therefore with a resolution of at most 2' —is being measured). When and where it is advisable to use the first or the second method? The answer is that the magnetic compass is a cheap, handy instrument that can always be used, in any weather conditions and in any situation. This means that it should *always* be used. The problem, rather, is knowing when it will be enough (Belmonte and Hoskin 2002). This would happen in cases where the geometry of the monuments cannot be ascertained with exactness (think for instance of the axis of a partly ruined megalithic tomb) and/or where there are hundreds of monuments to be studied in a limited amount of time (such as dozens of tombs in a necropolis). Equally, if the position of the monument allows a very clear determination of the direction that has to be measured—let us say, the sides of a Greek temple or the main urban axis of a town—then it is certainly wise to use at least one more refined instrument. This is particularly important if there is a strong likelihood that ancient surveyors were working with a very high degree of accuracy, as is the case, for instance, with the Giza pyramids (Sect. 8.1).

2.6   Virtual Globe Software

The use of *Virtual Globe Software* is of enormous assistance both in studying and researching archaeoastronomy. At the moment of writing, the most diffused free software of this kind available is Google Earth, on which I shall concentrate here (another very useful one is World Wind).

A virtual globe is essentially an extremely accurate computer version of a world globe. It maps out the Earth, area by area, by superimposing images obtained mainly from satellite imagery. It also provides 3D reconstructions of buildings and street-view navigation. The resolution is in most cases good and gives a fairly good picture of a site. Obviously, personal visits to sites are fundamental in archaeoastronomical research, but the program can be of great help to a student who wants to study, for instance, the urban layout of a Maya town without being able to fly off to visit it. The use of virtual globes is also recommended in archaeoastronomical research. Indeed, some sites may be inaccessible due to hazardous conditions prevailing locally, such as conflicts. What is more, landscape and visibility at many sites were different in ancient times, and these programs allow us to study visibility lines that are today lost or broken up by intervening obstacles. By way of example, as we shall see in Sect. 8.3, the main axis of the Necropolis of Giza points to the opposite bank of the Nile, towards a place (the Temple of Heliopolis) from which the huge pyramids of the plateau were plainly visible in ancient times. Today, alas, they are irrevocably hidden from sight owing to the encroachment of the buildings of modern Cairo as well as atmospheric pollution.

The tools an archaeoastronomer needs to use in Google Earth are simple:

(1) the *ruler* option, which allows the calculation of distance and azimuth between two fixed points.
(2) the *elevation profile* option which visualises the elevation between the two extrema of a chosen path and thus the projection of the line between the two points previously selected (Figs. 2.4 and 2.5).

The ruler thus gives the azimuth of desired directions: main axes of towns, sides of buildings, and the like. The elevation profile allows us to deduce the horizon height from a fixed observation point. A note of caution: the image shown—which

Fig. 2.4 Selinunte. Satellite view of Temple E, with the Google Earth ruler showing the azimuth of the building. Image courtesy Google Earth, drawing by the author

Fig. 2.5 Selinunte. Satellite view of Temple E, with the Google Earth ruler showing the azimuth of the building projected for several kilometers up to the hill at the local horizon (see Sect. 10.1 for details on this temple). Altitude profile is shown (Image courtesy Google Earth, drawing by the author)

is by default the one most recently acquired by the database—is not always the best one available for the area (for instance, a site might have been covered over for conservation reasons). To check previously archived images use the "show historical images" option.

It is a good idea to treat the program working space as an experimental field: repeat measures, always take them in both directions, and so on. If used correctly, in many cases an instrument as simple as the ruler gives such a good result that it can be used as an additional check of measurements obtained on field. Furthermore, it can be used to geo-rectify maps obtained by conventional means and/or found in old published works; it may frequently happen that such maps are badly oriented (that is, the north shown is not geographic north), not through negligence on the part of the surveyor but because magnetic north was used. A simple procedure allows us to superimpose the scanned image of the map in the program machinery and thus identify true north on the map with a reasonable approximation.

References

Aveni, A. F. (2001). *Skywatchers: A revised and updated version of skywatchers of Ancient Mexico.* Austin: University of Texas Press.

Belmonte, J. A., & Hoskin, M. (2002). *Reflejo del cosmos: Atlas arqueoastronómico del Mediterráneo Antiguo.* Madrid: Equipo Sirius.

Muller, J. (2012). *Geodesy.* London: de Gruyter.

Ruggles, C. L. N. (1999). *Astronomy in prehistoric Britain and Ireland.* New Haven & London: Yale University Press.

Chapter 3
Data Analysis

3.1 Reconstructing the Ancient Sky

Since all the objects in the sky obey to the deterministic laws of physics, their past and future movements are in principle predictable with the use of mathematical equations, especially for the exceedingly short (as compared to the life of the universe or even of the solar system) periods that are the focus of archaeoastronomy, let us say—to embrace comfortably the most ancient temples of humanity at Gobekli Tepe—the last 12,000 years. Traceability is also important for those few movements (the so-called "proper motions") that relate to single bright stars over the course of millennia (e.g. Sirius and Arcturus), which, if necessary, can be calculated with a reasonable degree of confidence. Up to the advent of personal computers, mathematical astronomy was a tedious chore, due to the difficulties of performing long equations by hand. Fortunately, today astronomical software does the job.

An astronomy software, or digital planetarium, is an instrument which calculates and consequently simulates the appearance of the sky at any chosen time; instead of projecting the sky on a dome, as in a real planetarium, it reproduces the sky as images on the computer screen. Such software applications are usually meant for use by astronomers equipped with (more or less powerful) telescopes, and therefore the programs' databases contain an enormous quantity of stars, usually hundreds of thousands. Of course, such catalogues are of no use in archaeoastronomy, and we should actually take care not to load them setting the upper limit to the magnitude of the objects displayed by the program to a reasonable human-eye threshold, say 5 (or at most 6). The software permits the choice of the observer's location (via geographical coordinates or by using names) and the time (including the year, typically for many millennia in both directions). With a digital planetarium we can simulate all phenomena of interest like sunrise and sunset on chosen days or the heliacal rising of stars. Constellations can be identified as boundaries and also as stick figures; sometimes pictorial representations are available. Atmospheric effects

© Springer Nature Switzerland AG 2020
G. Magli, *Archaeoastronomy*, Undergraduate Lecture Notes in Physics,
https://doi.org/10.1007/978-3-030-45147-9_3

Fig. 3.1 Planetarium simulation of the Heliacal rising of Sirius at Cairo in 2700 BC

can be simulated too. The most useful and relevant default options for archaeoastronomical purposes, besides the magnitude limit, are:

- a flat horizon (without panoramas), only outlined. In this way heliacal risings can be studied by checking the altitude of the Sun under the horizon
- azimuthal or polar grid
- equinoxes and solstices (when available) highlighted.

The best way to learn the use of these software resources (and to simultaneously acquire the rudiments of astronomy with the naked eye) is to *practise*, trying to reproduce with them the apparent motions of the celestial bodies described in Chap. 1 and solving the corresponding exercises at the end of the book (Fig. 3.1).

3.2   The Reconstruction of Visual Alignments and the Horizon Formula

As we shall soon learn, modern archaeoastronomy is not just the science of astronomical alignments but is a much wider-ranging discipline suitable for studying *any* kind of potentially intentional alignment between structures. Indeed, in many cases ancient monuments were interconnected with each other or with landscape features by visual lines that stemmed from symbolic considerations, forming sacred landscapes which fall well within the realm of archaeoastronomy.

Some such alignments are very clear and immediately recognisable, but some others are very long, and may be so long, in fact, that they are no longer effective today, due to pollution or other reasons. In any case, since the Earth is round, two points can be claimed to have been intentionally aligned only if they are (or were) inter-visible. Unfortunately, however, the web is full of alleged "alignments" between sites which can even stretch for hundreds of kilometres and are drawn as straight lines on the spherical surface of our planet…(as air pilots know very well, the only concept of a "straight line" between two points that has any sense on the Earth's surface is the arc of the maximal circle between the same points, because it is the shortest path; of course, ancient civilizations could hardly have developed such a concept on the ground). Fortunately, there exists a very simple mathematical tool which allows us to distinguish truly visual—and thus possibly non-coincidental —alignments from all the rest. It is the Horizon Formula, which is actually a straightforward follow-on of the Pythagorean theorem.

Given a point P located at vertical height h with respect to the Earth surface, the farthermost point W that can theoretically be seen from P of course is the point where it passes the tangent to the Earth circumference from P. Clearly then the distance d between P and W satisfies to $d^2 = (r+h)^2 - r^2$ so that $d = \sqrt{(2rh + h^2)}$ where r is the Earth radius. If the heights are very small with respect to r we can safely discard h^2 (to see this, rewrite as $d = \sqrt{2rh(1 + h/2r)}$ and observe that h/2r is much smaller than 1) so that by all practical means it is $d = \sqrt{(2rh)}$. Since $2r \sim 13{,}000$ km, we arrive at the above mentioned "horizon formula": *the visible horizon d in kilometres equals with fair approximation the square root of 13 h if h is expressed in metres* (Fig. 3.2).

This is a very handy, easy-to-use formula. It shows, for instance, that a person 2 m tall has a visible horizon $d \sim \sqrt{26}$ km, that is slightly greater than 5 km, while a Maya pyramid 70 m tall (as is the so-called "Temple 4" at Tikal, Guatemala)

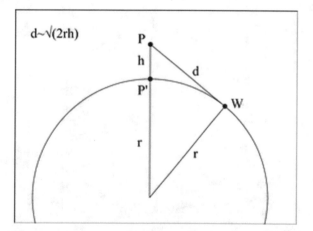

Fig. 3.2 The geometry needed for the horizon formula. Given a point P located at vertical height h with respect to the earth surface P', the farthermost point W that can be seen from P is the point where passes the tangent to the earth circumference from P. The horizon formula is an estimate of the distance d = PW valid when h is small with respect to the earth radius

is theoretically visible from a distance $d \sim \sqrt{910}$ km so from about 30 km (in case of two tall objects, the heights—not the distances!—must be added). It should, however, be noted that local atmospheric conditions may reduce visibility significantly.

3.3 Graphical Tools

Using an astronomy software application, one can identify the celestial phenomena the builders of a monument in question might have been interested in. To compare the results with the experimental data (azimuths and horizon heights) there are two particularly simple instruments: orientation diagrams and histograms (Fig. 3.3).

The orientation diagram is a simple graphical tool, typical of archaeoastronomy studies, which can be used as a first approach in dealing with a set of data. To obtain an orientation diagram, draw a circle and bisect it with two orthogonal lines parallel to the sides of the sheet (or the virtual sheet on a computer screen). These will be the south-north (north on top) and east-west lines. Now from the centre draw a line level with each azimuth you have found on field. If the horizon is flat or nearly flat, the result will provide quite decent, albeit only indicative, information about the behaviour of the data which, if the orientation is not random, will be seen to concentrate in specific sectors. In many cases, it may also be useful to draw the solstitial axes in the diagram, highlighting them with a different colour or a type of drawing; other azimuths of astronomical interest (of the Moon, or of bright stars, or whatever) can be highlighted as well.

A histogram is a standard graphical instrument which helps in assessing the presence of non-casual peaks in the distribution of a set of experimental data (Fig. 3.4).

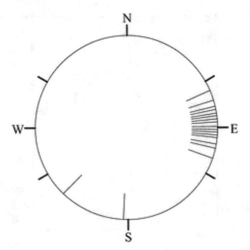

Fig. 3.3 An example of the use of orientation diagrams in archaeoastronomy: the orientation of 28 Greek temples of Sicily. Only two azimuths fall outside the solar arc of the Sun at rising (data from Aveni and Romano 2001)

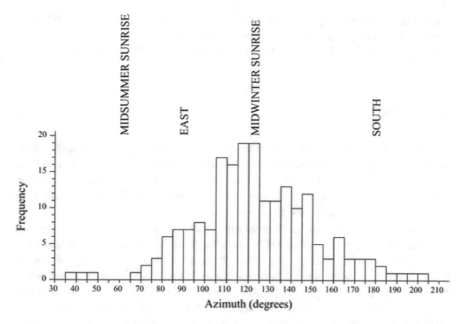

Fig. 3.4 An example of the use of histograms in archaeoastronomy: azimuth histogram of the orientation of 210 megalithic sepulchres in Almeria, Spain. Almost all the data fall in the sector of the sun rising (the solar arc, between the two solstices rising azimuths) or climbing in the sky (between the winter solstice rising azimuth and due south) (data from Hoskin 1998)

In archaeoastronomy, one can use azimuth or declination histograms. To construct a histogram, the first step is to divide the range of the possible values into boxes—or intervals—of fixed dimensions. If we want to construct a azimuth histogram, the full range of 360° can be conveniently divided into, say, boxes of 5° or less. Then, the number of values that fall into each interval is counted. A rectangle is drawn with height proportional to the count and width equal to the size of the boxes. Another possibility is to "normalise" the data, dividing each item by the total number and therefore displaying relative frequencies. However in archaeoastronomy the quantity of data is usually small, so I do not recommend carrying out this normalisation since it may conceal the fact that peaks may have actually been generated by too small a number of matches.

Both the orientation diagram and the azimuth histogram may have serious disadvantages. The first is the fact that the horizon at the site being studied may be uneven, and therefore azimuths referring to the same astronomical event may be different; the second (which may be tied up with the first) intervenes if data from different sites are put together. If indeed the latitude differs too much, the azimuths of the same astronomical events will also differ. The solution is to switch the data into declinations and use declination histograms—declinations take into account both horizon and latitude, giving an "absolute" result. Of course, at a given site only a restricted range of values of declinations correspond to the whole visible sky, and this range has to be calculated before putting the data on the histogram.

3.4   Statistical Tools

The statistical analysis of experimental data is a fundamental task in any applied science. Statistics can—and must—be similarly applied to sets of archaeoastronomy data. However, as we shall see, the cultural/sociological nature of data here means that we have to apply statistics in archaeoastronomy in a very prudent and responsible way (Ruggles 1999).

Let us suppose to have a building which exhibits a certain number of alignments (e.g. different windows), or a set of buildings in which we have measured a potentially interesting feature (e.g. the entrance corridors of a group of tombs). A first note of warning must be sounded immediately. Statistics is an extremely evolved science that provides a series of complex and sophisticated mathematical instruments for the purposes of data analysis. And yet, in archaeoastronomy, it is difficult to have datasets with more than a few dozen values; as a consequence, it is advisable to use rather elementary statistical approaches as the one I am about to describe here. A second note of caution. Sometimes, an effect called *data selection* can undermine any statistical reasoning right from the outset. To avoid data selection one must take great care in measuring and considering any possible alignment at a site (for instance, between stones in a megalithic monument), not just choosing those which are likely to have an astronomical interpretation. Similarly, one should preferably decide a priori which astronomical targets might have been of interest for the builders, and not a posteriori. It is, in fact, easy to match a certain azimuth with a star, especially if one juggles around with the allowed error, or with uncertainties in the date of construction, or with both (see also Sect. 6.2). All in all, the definition of targets of potential interest must be based on reasonable cultural arguments. A (non-exhaustive) list of targets to be considered as eligible might be the following:

- the four cardinal directions
- the two solstitial axes
- the arc of the Sun between the solstices
- the arc of the Sun climbing in the sky (from the azimuth of the winter solstice to due south)
- the major lunar standstills
- the rising/setting of brightest stars, at least in all cases in which a cultural interest in them is soundly documented (for instance in ancient Egypt) and a reasonable date for construction is known
- slightly less obvious targets, if there are historical/cultural reasons for them.

Examples: the Zenith passages of the Sun for the pre-Columbian cultures of Mesoamerica, the Pleiades for the Incas; Venus for the Mesoamerican cultures; the Pleiades and also faint but culturally relevant constellations such as Delphinus for the Greeks, and so on.

Once a satisfactory definition of potentially interesting targets has been agreed on, we have still to agree on the allowed error, in other words, we have to define what matching targets means. This is quite a delicate point, since there are two completely different things—confusingly, with the same name—going on. The first is the *modern* error, the one which is intrinsic to our measurements. If we used a magnetic compass, it can be quite substantial, as we have seen (not less than $\pm\frac{1}{2}°$). So if the azimuth matches the target within the modern error, it, of course, matches. But another error, which we cannot really foresee or control, is present: the original one. As a matter of fact, ancient surveyors also used an instrument (the naked eye, backed up with natural or artificial foresights) and therefore committed an error. It is thus usually necessary to have an idea of how much the original error might have been. For instance, for observations made on the flat horizon of the jungle in the tropical area of the Mayas we would not expect the original error to have been less than $\pm 1°$, while we have incontrovertible proof that Roman surveyors were able to measure the meridian within $10'$, and that the ancient Egyptians oriented the pyramids of Khufu and Khafra at Giza with even greater precision, to the point that measuring the side of the Khufu pyramid with a magnetic compass would be useful only if we want to use the measure to have an estimate of the error of the compass we are using....

Once all of this has been carefully considered can we finally seek an answer to the question: are the astronomical correspondences that have been found merely fortuitous, or do they reflect the explicit intentions of the builders? To solve this problem we can use probability. To start with, let us consider a marksman who fires randomly n times, without looking. What are the odds of him hitting exactly k targets? If p is the probability of hitting a target with a single shot, probability P is a function of n and k given by the Binomial distribution

$$P = \frac{n!}{k!(n-k)!}\, p^k (1-p)^{n-k}$$

where the "factorial" symbol! after an integer means that the product of all positive integers less than or equal to that number must be taken, for instance, $4! = 4 \times 3 \times 2 \times 1 = 24$.

To apply binomial distribution in archaeoastronomy, suppose that in measuring a site, out of n total alignments, we have found k pointing to astronomical targets. What are the odds that this result has come about by chance? To calculate this we must first estimate the probability p of a single alignment occurring. Then, we can use the binomial distribution formula to estimate what the probability is that the result—k out of n—is down to a chance, and express the result in the form of a fraction. For example, we may find that there is a 1/1000 probability of getting the result by chance, or 3/100, or 1/25, or whatever. The smaller these fractions are, the higher the chances for *non*-coincidental orientations (see e.g. Polcaro and Polcaro 2006).

Of course, naturally, if we had found a number of alignments *greater* than k, we would have tried to interpret this information in the same way. This means that we are not quite approaching the issue in a correct manner, as we should be considering the probability of reaching k or more targets, not *exactly* that. Clearly, this *cumulative* distribution will always give greater probabilities as compared to simple distribution. By adding up the single probabilities appropriately, such a distribution can be calculated relatively easily. For the sake of simplicity, I use here only the standard binomial formula; the results actually do not differ qualitatively each time the single shot probability is significantly less than, say, ½, as usually occurs in archaeoastronomical applications.

Imagine now that a chance probability has been obtained, and that this probability is "small". We have two problems. First of all, even if we accept that the orientations are definitely not coincidental, the astronomical interpretation might not be correct. A trivial example might be that of a cluster of tombs excavated in a rock wall. The common orientation clearly is roughly orthogonal to the wall, for geo-morphological reasons, but it may, by chance, coincide with an astronomical target. A much less banal example would be that of buildings orientated towards natural features of the landscape, such as, for instance, a mountain. In the first example, our analysis if no use, as the alignment is due to geo-morphological reasons, but in the latter it may reveal that the mountain was sacred, and deliberately selected.

Another issue is whether we have any way to give a mathematical meaning to the concept of "small" probability. For instance, to what extent are we prepared to believe in an astronomical hypothesis if the probability is 1/1000, or 3/100, or 1/25? The answer must come from the establishment of a common agreement on the threshold which allows us to claim that a phenomenon is not random. Since this problem is shared by all the sciences, it is certainly not the job of archaeoastronomy to establish this threshold, but, rather, we should conform to the methods commonly adopted in all the other sciences. This method is based on the obvious fact that the further a result deviates from the mean, the greater the chance it will be statistically significant. To measure this distance the concept of *standard deviation* can be introduced.

For the binomial distribution the mean is given by the product np between the number of cases and the single probability, and the standard deviation is

$$\sigma = \sqrt{np(1 - p)}$$

It can be shown that 68% of the values of a binomial distribution fall within one standard deviation from the mean, 95.5% within two standard deviations from the mean, and that as many as 99.7% fall within three standard deviations. Therefore, a common—and quite natural—convention is to assume that data start to be significant if they are at least "at the 2σ level"—that is, their distance from the mean is greater than 2σ—and that they are at a high level of confidence if they pass 3σ.

The threshold at 2σ is, however, by no means universal: for instance, C14 dating used in archaeology is often given only at a worrying 1σ level of confidence, while in frontier experiments in fundamental physics, levels of confidence of 5σ are sometimes required. What about archaeoastronomy? As a matter of fact, seeking a 3σ level by default (as some would advocate, see e.g. Schaefer 2006) is simply a way of destroying the credibility of much of this discipline, given that the data samples are usually very limited. For example, it can easily be seen that any assertion of intentional alignment at a single site (like a row of megaliths) would thus be automatically rejected. Indeed, even considering only the solstitial and the cardinal directions as significant, one has eight possible azimuths which, allowing $1°$ of error, give a probability for a chance alignment along one of them of 4.4% (since $16/360 = 0.044$), which is just below 2σ level and very far from the 3σ. Clearly then, such a drastic approach cannot be accepted. Actually—and amazingly —what happens in many cases is that only alignments which would have no need for the use of statistics to be considered as intentional by any sane-of-mind person can be claimed to be accurate at the 3σ level (we shall see two examples in this book: the claims that the Greek temples of Sicily are oriented towards the rising Sun, and that the megalithic sepulchres of Iberia are oriented within the arc of the Sun's rising/climbing).

To conclude, the statistical approach must always be considered with reference to the special characteristics of archaeoastronomical data: a physicist who is unconvinced of his experimental results at 2σ level can repeat the experiment to improve statistical analysis, but we cannot erect new megalithic tombs to extend our samples. Of course, on the other hand, iniquitous use of statistics through data selection, target selection and the like must be strenuously avoided. Finally, as far as single sites are concerned, I personally would argue for the right for archaeoastronomy to study single, unique sites and to come up with some very reasonable conclusions, with or without any σ level that might be applicable. We shall see examples of such one-off sites in this book, and three of them are—I repeat, at least in my view—among the most outstanding successes of archaeoastronomy in absolute, as well as being among the most beautiful and complex monuments ever created by the human mind: the Pyramid of Khufu at Giza, the Castillo at Chichen Itza' and the Pantheon in Rome.

References

Aveni, A. F., & Romano, G. (2001). Temple orientation in Magna Grecia and Sicily. *Journal for the History of Astronomy, 31,* 52–57.

Hoskin, M. (1998). Studies in Iberian archaeoastronomy: (5) orientations of Megalithic Tombs of Northern and Western Iberia. *Journal for the History of Astronomy, Archaeoastronomy Supplement, 29,* S39.

Polcaro, A., & Polcaro, V. (2006). Early bronze age dolmens in Jordan and their orientations. *Mediterranean Archaeology and Archaeometry, 6,* 169–174.

Ruggles, C. L. N. (1999). *Astronomy in prehistoric Britain and Ireland*. New Haven & London: Yale University Press.

Schaefer, B. (2006). Case studies of the three most famous claimed archaeoastronomical alignments in North America. In T. Bostwick & B. Bates (Eds.), *Viewing the Sky Through Past and Present Cultures: Selected Papers from the Oxford VII International Conference on Archaeoastronomy* (pp. 27–56). Phoenix: City of Phoenix Parks and Recreation Department.

Part II
Ideas

Chapter 4
Astronomy and Architecture at the Roots of Civilization

4.1 From Homo Sapiens to Homo Sapiens

Nowadays most of us do not look at the sky.

Nevertheless, most people know "what star sign they belong to". And yet the vast majority of people do *not* know what "belonging to a zodiac sign" actually means and have never seen their "sign" in the night sky. Indeed, hardly anyone ever glances at more than a few stars at night—for most people the issue simply never arises.

Once upon a time, it was not so.

People did look at the sky, and the sky appeared much more brilliant and densely populated than it does today, viewed from almost anywhere on earth. This, one might surmise, is due to modern pollution. Modern pollution is indeed a problem, but—leaving aside the cases of some extremely polluted cities—this is not the greatest problem. The chief one is *light* pollution: the presence of ubiquitous, disproportionate artificial light. To understand how light pollution affects our view of the sky, we need only think that even the light from one street lamp can destroy the view of the Milky Way, or that the lights of a distant town on the horizon will usually restrict the visibility of the stars less brilliant than the small number of 1-magnitude ones. So, if we want to see the sky as it was before the spread of large-scale electric illumination we have to go to very special places: isolated—like mountain summits or, even better, mountain highlands with a high horizon all around—or places kept in the dark for particular reasons (like, for instance, the Akumal beach in Mexico, where lights are turned red and kept to a minimum to facilitate turtle nesting). In any case, there is nothing—believe me, nothing— guaranteed to gladden a skywatcher's heart like a total blackout with a clear sky.

So the ancient sky was once an overwhelming, dominant presence at night. Thousands of stars could be seen, and a celestial river flowing with the very same velocity as each star turnaround occupied a wide band all along the firmament. But, apart from the aesthetic aspect, why and when did people start to look at the sky?

© Springer Nature Switzerland AG 2020
G. Magli, *Archaeoastronomy*, Undergraduate Lecture Notes in Physics,
https://doi.org/10.1007/978-3-030-45147-9_4

The answer is far from simple. The great Roman poet Virgil once wrote that humans became really human only when they started counting the stars in the sky, and perhaps he was right. But to obtain more insight into this, it is first necessary to recall a few fundamental points in human history.

People defined as "anatomically modern humans", that is, *Homo Sapiens*, are attested to in Africa from around 200,000 years ago. As far as we can judge from material remains, their intellectual behaviour underwent a profound change at the beginning of the period defined as Upper Palaeolithic, which, very broadly, spans the years between 50,000 and 10,000 BC. It is certain that the first half of the Upper Palaeolithic marked some kind of breakthrough in human history (notice that the use of the words "human evolution" would be completely out of context here, since we are speaking of our own species). Archaeological records show two fundamental innovations. The first was a process of geographical expansion, triggered off for (so far) unexplained reasons, which led humans to settle in Europe during the millennia around 40,000 BC. Europe was at that time inhabited by another human species, the Neanderthals. There is absolutely no proof whatsoever that these humans were "less evolved" than we were. Rapidly, however, and for no clear reason, the Neanderthals became extinct. This happened precisely during the millennia following the arrival of the Sapiens, so that one possibility, which appears to be supported by genetic evidence, suggests that the two species were fertile and that the Neanderthals actually merged with the Sapiens, thus contributing to our present DNA.

The second breakthrough, probably commencing previously in Africa before the expansion, was an explosion of technical and artistic skills (Mithen 1998). It was, firstly, a sort of technical revolution, documented archaeologically in the division of stone artefacts into dedicated instruments (engraving tools, knife blades, and so on). Secondly, it was an explosion of a variety of different arts. For instance, refined statuary in stone, ivory and bone appeared. The art of painting was invented, apparently from scratch, by artistic geniuses who started to decorate the interior of caves with (mostly naturalistic) images. Thirdly, pottery was invented with a purely symbolic aim—that of creating female statuettes, presumably of divinities.

We do not know the reasons for such a boom, which might be defined as a cognitive revolution; some have tentatively suggested that it was in this period (and not before) that abstract language appeared, so that the increasing complexity apparent in the archaeological data we have stems from a corresponding complexity in human-human interaction with regard to concepts. Others reject as unacceptable the sudden appearance of "modern" behaviour, contending that only a millennia-long process—though this is not supported in the records—could produce this, invoking thus a linear concept in human development. This theory appears unsound, however, also judging from a number of other historical considerations. In fact, up to the first half of the nineteenth century, the development of new human capacities was always deemed to be what a physicist would call a *linear process*. A trial-and-error succession of approaches which progressively leads to improvement, without any recognizable break, step, or leap, and—often—without recognisable cause-effect mechanisms. For instance, the numerous depictions of human hands,

Fig. 4.1 Lascaux. The hall of the bulls in the (reconstructed) Lascaux cave

obtained by pressing freshly-painted hands onto the walls of Palaeolithic caves, were considered to be the first attempts at "art", obviously preceding the painting masterpieces of world-famous sites such as Lascaux or Altamira (Lawson 2012) (Fig. 4.1).

Similarly, the invention of pottery was considered as having a functional use in agriculture (which, as we shall see, came later), and therefore was not really possible in the Upper Palaeolithic period.

Then a series of recent discoveries, together with the development of Carbon Dating, shattered this view of human sensibility in the Palaeolithic. Let us start by considering cave paintings. In the spring of 1994, the archaeologist Jean-Marie Chauvet, exploring the Ardeche Gorge in south-eastern France in the company of some friends, discovered a previously unknown frescoed cave. The artist or artists who worked there painted more than four hundred figures of animals: mostly lions, horses, oxen and rhinoceroses. The pictorial technique is splendid and shows a firm grasp of perspective. So it would be logical to attribute the cave to the same period as the other known masterpieces of cave art, say around 15,000 BC. However, incredibly, carbon samples from the Chauvet cave yielded dates of around 30,000 BC. Consequently, any idea of the snail's pace development of artistic capacities was now to be discarded. The reason is simple: human thought is non-linear, and does not necessarily require long drawn out processes to achieve outstanding results. Furthermore, invention does not necessarily have to be stem from purely functional considerations. Here the key example is, as mentioned above, pottery.

A simple, reassuring way of understanding the development and the widespread use of pottery was to attribute it to the functional needs of the first sedentarists (who needed it to store seeds in). Today we know that this is categorically untrue, as pottery vessels are documented in China thousands of years before the birth of agriculture, and were created in many places (like the Jomon civilization of Japan) around 10,000 BC, if not before. But this is not far enough back. Pottery was actually invented in the Upper Palaeolithic Age. The first examples of baked pottery are female statuettes from Dolní Věstonice in modern-day Czech Republic, dated to the period 29,000–25,000 BC.

So, it can be seen that any attempt at constructing a gradual evolution model to explain the astonishing progress of Homo Sapiens is doomed to failure. It is much better to accept that human thought is ultimately nothing more and nothing less than a very complex physical process, and can therefore proceed non-linearly. So, a linear progression *can*—but need not—be at the basis of a new idea, invention, thought, or even an entire science. This is probably also true for astronomy, which seems to have been born—together with arithmetic—in the Upper Palaeolithic as well (Rappenglück 2008, 2011). A first, albeit controversial, hint is given by the possible presence of calendars in artefacts of the period. This has been the subject of a wide debate since Marshack (1972) identified series of notches on incised bones as calendrical marks connected with a lunar count. Among these, in particular, a bone dated to about 28,000 BC from Abri Blanchard (Dordogne, France), where the dots are interpreted as a count of successive phases of the moon. Marshack's work has been much criticised on the grounds that the markings did not accumulate on subsequent days, but were only made on few occasions (D'Errico 1989). It is, however, also true that someone may have incised the marks representing, rather than counting, a cycle; in any case, the evidence is not conclusive and a complete re-analysis of the available records might be worthwhile (D'Errico and Cacho 1994; Rappenglück 2012).

The situation becomes clearer if we look for proof, in the same period, of a less specialised interest in time, limited perhaps to the seasonal sequence. Seasons are indeed signalled by images of specific animals carved on bones, and animals relating to seasonality and time are also frequently found in cave art. Some of such depictions appear to have an "almanac" content, with different animals associated with different seasons, and dots perhaps representing calendar counts. Furthermore, some artworks might be interpreted as depictions of asterisms in the sky (Rappenglück 1998). In particular, in the so-called Hall of Bulls of Lascaux, the depiction of a bull with six dots arranged in two rows above the shoulder and other dots surrounding the head as a V-shape clearly resembles our way of representing the constellation Taurus (the six dots would be the Pleiades). Similar constellation-like images are found in other caves, including Tête-Du-Lion (Ardeche, France). Of course, to be absolutely certain of the astronomical significance of this kind of depiction it is necessary to presuppose cognitive continuity in the perception of the sky as well as in the popular imagination (Pásztor 2014). However, some astronomical interpretations of Palaeolithic cave art can be placed within a fairly sound cultural framework. This framework is that of *shamanistic*

Fig. 4.2 Lascaux. The bird-man fresco in the (reconstructed) Lascaux cave

cosmology. As we shall see, we have clear indications that the religion of the Upper Palaeolithic and of the Neolithic was of shamanistic type (Sect. 5.3) and we know a great deal about shamanism from contemporary anthropology. As a convincing example of shamanistic interpretation we might consider the famous and enigmatic painting which is located in a small room, or shaft, of the Lascaux cave (Fig. 4.2).

The Lascaux cave has many features which point to its use as a place where people gathered for cults, especially the magnificent frescoed Hall of the Bulls, mentioned previously. The shaft is a recess, a sort of holy of holies not easily accessed. Here the frescoed wall shows a therianthrope, a half-man/half-animal creature with the head of a bird and body of a man, with erect phallus. The right hand of the bird-man seems to be holding a stick, though there is no contact between hand and stick. The handle of the stick is also carved in the form of a bird, perhaps a dove. To the bird-man's left, a large bison lies dying, shot full of arrows. A woolly rhinoceros and a horse complete the scene. To interpret this enigmatic depiction as a representation of the sky, we must recall that—because of precession—the sky above the painters of the Lascaux cave was very different from ours. In those days, the north celestial pole was crossing the Milky Way. It was not close enough to any "pole" star, but it was not far from Delta Cygni, and so the great bird of the sky was seen to rotate closely tied to the pole. The Bird Man might thus be a celestial figure stretched across the Milky Way, the upper part constituted by stars from those constellations we call Cygnus and Vulpecula, while the lower part includes stars from our Aquila, Serpens, Hercules and Sagitta. The bird-headed staff may represent the polar axis (Rappenglück 1998). If this is accepted, then the other figures also come to life as constellations: the three animals that surround the scene—a bison, a woolly rhinoceros and a horse—are indeed recognisable as three large constellations associated, respectively, with the north-west (autumn), north-east (spring) and south, with the horse corresponding to the constellation we call Leo. The sky depicted in

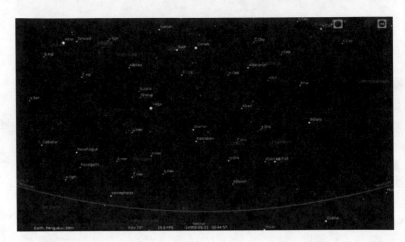

Fig. 4.3 Lascaux. The region of the North celestial pole as it appeared around 15,000 BC. The pole is close to the star delta-Cygni

this way is likely to be that visible from the top of the Lascaux hill, around midnight, at the summer solstice in the years around 14,500 BC (Fig. 4.3).

The shamanistic interpretation of this scene is suggested because, as we shall see in the next chapter, a fundamental duty of shamans is to act as intermediaries, putting humans into communication with the divine through a "cosmic" voyage in a trance state. Shamans usually hold sticks, their iconography is often related to birds and in some cases they even claim to be able to transform themselves into birds during trance. The scene thus fits in neatly with the representation of the cosmic voyage of a shaman.

To conclude, although single cases might be disputed and definitive evidence is still lacking, it is being increasingly suggested that there was considerable focus on the sky during the Upper Palaeolithic period. Perhaps even the choice of specific caves might have been inspired by astronomical considerations, as for instance for the Cova del Parpalló, one of the most important Palaeolithic Sites of the Mediterranean coastline of the Iberian Peninsula, where the innermost chamber of the cave marks the coming of the vernal equinox (Lull 2014). Such a focus might well have arisen for *functional* reasons: the presence/absence of moonlight was needed for nocturnal activities, the cycle of the seasons was needed to follow the migration of animals and maturation of various fruits, the stars were needed for orientation purposes, and so on. However, it is likely that the *symbolic* world of these people also included references to the sky.

This becomes abundantly clear when we move a little closer to our times, and pass to that game-changer in human history: the invention of agriculture.

4.2   From Hunters-Gatherers to Herders-Peasants

The birth of agriculture is placed chronologically at the end of the Upper Palaeolithic period. It is of fundamental importance here to specify what kind of human society existed in those times. All human groups living in the Upper Palaeolithic are unanimously defined as "hunter-gatherer" societies: their consumption of flesh derived from hunting, their consumption of vegetables derived from gathering.

It is essential to stress that the term hunter-gatherer must *not* be taken to mean "primitive", although such a meaning is often bandied about and accepted. Actually the definition "hunter-gatherer" was coined to distinguish between human groups of people who live without fixed, stationary dwellings and sedentarist groups of people. In fact this terminology was actually drawn up by sedentarist people (yes, us!). It would thus be quite fair to denote ourselves with an equivalent term— namely "herder-peasants". Sources of human food in fact have not really changed since prehistory, and we can have the most industrialised society in the world complete with international space station and worldwide web, but we still depend on farming and herding. This should remind us, for once and for all, that there is no reason to suppose that the brains or intellectual abilities of hunter-gatherers were inferior to ours. And indeed, a hunter-gatherer does not hunt and devour any old beast that stumbles into his line of his sight, but must develop hunting techniques and be an expert in animal behaviour, reproduction, and migration. Likewise a hunter-gatherer does not gather and devour any old edible thing he comes across, but must become an expert in the cycle of seasons and of the growth periods and location of edible plants and fruits: hunter-gatherers were well-versed in natural history (Lewis-Williams and Pearce 2005). Hunter-gatherers were the people who first developed art, knowledge and religion long before the invention of agriculture, and these were the people who, as we shall shortly see, developed the first monumental architecture and—probably—also the idea of connecting this architecture with the sky. For these reasons, the term hunter-gatherer will no longer be used in this book.

The process relating to the introduction of agriculture is usually referred to as the Neolithic revolution. Humans began to form farming communities and to establish permanent settlements around 8000 BC. A process was set in motion which rapidly induced humans to become sedentary through the invention and introduction of agricultural techniques and the domestication of animals. Our understanding of the details of these processes today is, however, rather vague and nebulous compared to what was considered gospel only a few decades ago (Runciman 2001). Indeed, rigid diffusionist ideas pervaded almost all fields of ancient history study. There existed— it was thought—a sort of perennial cradle of civilization in the near east. From there a revolution bringing magic discoveries—splendid agricultural sedentarism and the wonderful techniques of animal domestication (seen as vast improvements in the quality of human life)—spread in every direction. A clear parallel was made with the idea that, several millennia later, a cultural revolution bringing the advances of the Sumerian civilisation—for instance, writing and elaborate architecture—allegedly spread from Mesopotamia throughout brutal, barbarian Europe. In particular, megalithic monuments like Stonehenge were thought of as reflections of the Bronze Age Mycenaean culture of Greece, itself in turn influenced by oriental enlightenment.

Such an idea was out by some two millennia. Today we do know that cultural diffusionism did not occur, and many—myself included—also think that the above mentioned ideas on a similar diffusion of the Neolithic revolution have to be radically reconsidered. This reassessment is necessary both from a technical point of view—as many hints show, for instance, that the agricultural revolution spread too rapidly in some areas to be considered a real diffusion, while delayed millennia in others, including Neolithic northern Europe, where only pastoralism is really documented—but also, and this is our main focus here, from a cognitive point of view.

To understand this, we start from a question whose answer has been taken for granted for decades. Are we really sure that sedentarian life is better? The work of the farmer is hard, and life is subject to the whim of nature at least equally, if not more so, than the life of the gatherer. Domesticated plants are vulnerable to the spread of diseases, and crops are sensitive to rain regimes. Fields can exhaust their harvesting potential, and the development of hydraulic and irrigation systems is usually a necessity. In actual fact, we can by no means be sure that this revolution perhaps was nothing more than—yes—*an error*. Perhaps, however, the error was not a real choice: climate change and population growth may well have influenced it. Yet, there is another possibility. As we shall shortly see, we do know that the formation of a complex system of religion and symbolic beliefs (connected to, but probably somewhat different from, those of the Palaeolithic period) occurred precisely during those millennia. So it may be, as is usually assumed, that this system was developed *as a consequence* of the dramatic change in the structure of the society brought about by sedentarism.

But, then again, it may well be the opposite.

In other words, there are indications that lead us to think that the stabilisation of human nuclei occurred, at least in some cases, prior to the adoption of agriculture. So what triggered the need for local production, herding and sustenance might have been a choice arising from sedentarism, and not vice versa. Agriculture and pastoralism may just be a by-product. But a by-product of what? The answer might be at once childishly simple and breathtakingly complex: religion (Cauvin 2000). A possible alternative view thus is that the revolution of the Neolithic period was primarily a symbolic one. The emergence of collective cults and behaviour, the birth of cult centres of attraction and pilgrimage and the foundation and spread of cult practices based on animal sacrifice, might have been the real impetus that prompted the formation of more aggregated societies, which were concentrated close to ceremonial centres and needed local means of sustenance and domesticated animals for food and offerings.

A very similar scenario can be posited for the development of astronomy, and this is why this issue is so crucial for us here. One might in fact think that sedentarism led to the need for a precise calendar for agricultural timing, and this in turn prompted astronomical observations. But one might equally think that the development of a complex religion, strictly connected with the celestial cycles by means of symbolic icons and the management of temporal power, led to the development of astronomical knowledge.

The keys to these questions are perhaps still partly buried in some special places which we are now about to visit.

4.3   The Birth of Monumental Architecture

Up to a few years ago, after the odd ideas about diffusionism had been consigned to the dustbin of history, the beginning of monumental architecture was thought to have virtually coincided with the onset of megalithic culture in Europe. In this book, we shall examine in detail some of the masterpieces of this culture, such as Stonehenge, the monuments of the Bru na Boinne valley, and others (Chap. 7). The age of such monuments is prodigious, but if we go back to the end of the fourth millennium BC, we find the existence of even older megalithic monuments (Fig. 4.4).

Fig. 4.4 Malta. The megalithic temple of Mnajdra ⊙

First of all, there are the temples built on the islands of Malta and Gozo (Trump 2002). The construction of these buildings began in the so called Ggantija phase (3600–3200 BC) and continued up to 2500 BC, before ending as abruptly as it had begun. The exterior of these temples presents a monumental portal façade, while the interiors are based on a architectural unit which is replicated two or more times; in many cases additions based on the same unit were made over the centuries. The unit is based on twin lobes, with a central axis usually perpendicular to the entrance. A huge exterior wall encircles the building (sometimes entailing the use of hundreds of tons of material) masking the complexity of the interior. The sequence of apses can vary but suggests a modular evolution. In some cases, however, entire new temples—albeit almost identical in layout—were built alongside the pre-existing ones, with changes in the orientation. For example, this occurs at Ggantija, where the temple complex comprises two very similar structures built at a distance of several centuries from one another. Here the basic module is a trilobate cloverleaf form, with an ovular entryway that leads inside by way of a small corridor delimited by giant facing stones. A similar example is Mnajdra, near the sea, while not far off, but located uphill, the temple of Hagar Qim shows accumulated additions to the same edifice (Fig. 4.5).

The guiding logic behind these buildings is truly unique, and at a first glance seems to elude any attempt at interpretation. The key must be sought in the cults which were practised there. They are represented in particular by a number of

Fig. 4.5 Malta. The megalithic temple of Hagar Qim

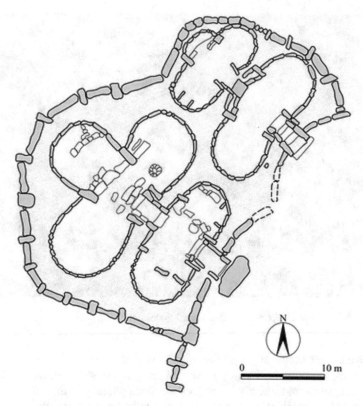

Fig. 4.6 Plan of the two Ggantija temple, Malta ⊙

statuettes with highly exaggerated, opulent feminine features, and by a huge statue of the same type. Such images have been ubiquitous in the Mediterranean area since the beginning of the Palaeolithic Age, and are related to the goddess usually identified as Mother Earth or Mother Goddess (Gimbutas 2001). Unfortunately, we do not understand much of the symbolism and of the rites associated with her, although there were undoubtedly connections with fertility rites, as well as with the sky, as we shall see in next section. Traces of what were probably animal sacrifice rituals have been found in the temples, and archaeologists have discovered further evidence of religious activity in the so-called oracular windows, round holes cut through certain wall slabs which are thought to have been used for oracular and confessional purposes.

The temples are therefore interpreted as shrines, or oracular shrines, to the Mother Goddess. Although not all archaeologists share this view, the iconography of the Goddess is the most likely key to the plan of the temples. In fact, the meaning of the internal lobed layout becomes obvious when one realises that it is none other than the silhouette of the body of the Goddess, seen from above. The temples are types of giant replicas of the divinity, following a symbolic mechanism that we shall discuss in general terms in the next chapter (Figs. 4.6 and 4.7).

Fig. 4.7 Malta. Statuette representing two seated goddesses

An even older group of megalithic monuments than the Malta temples is to be found, again, in continental Europe. For instance, in the area of Carnac (Brittany), megalithic activity was already in progress at the beginning of the fifth millennium BC. The oldest megalithic structures of Brittany are passage graves, like the tumulus tomb of Kerkado, Carnac, dated to about 4800 BC. Equally old are some stone circles in Portugal, such as the impressive oval of Almendres (Figs. 4.8, 4.9 and 4.10).

We are thus slowly moving back in time towards the limit of the Neolithic "revolution". However, all the structures we have described so far were built by sedentarist people, *after* that agricultural style had taken firm hold in Europe. Since these megalithic buildings were considered to be the oldest structures ever, as is natural, the linear idea comforted historians with regard to the birth of monumental architecture: that people *first* became sedentarist farmers, *then* started aggregating in stable settlements and societies, and finally these societies started to build monuments.

This theory is wrong, as evidence that humans were able to conceive of and build complex structures long before becoming sedentarists is today overwhelming.

First of all, some very ancient structures, albeit not built in stone, have been documented in central Europe. These structures (generically called circular enclosures although only a few are strictly circular) are made up of one (or more, concentric) earthwork separated by causeways and enclosing wood buildings, of which the postholes remain. There are more than 100 such enclosures, and others are being discovered by means of aerial photography. Perhaps the best known is the

Fig. 4.8 Carnac. A view of the alignments at Kermario

Fig. 4.9 Carnac. The tumulus of Kerkado

so-called Goseck circle in Saxony-Anhalt, Germany. Gosek appears to have been in use for a relatively short period between 4900 and 4700 BC. It consists of concentric ditches which were inter-walled by palisade rings; there were three openings

Fig. 4.10 Almendres. The stone circle

allowing access to the site. The function of the structure was probably ritual; excavations yielded what seem to be the traces of such rituals and perhaps also of human sacrifice. It is likely that even more ancient structures will be come to light in future: one example is the excavation at Warren Field, in the River Dee valley of Scotland's Aberdeenshire, where a series of postholes dated to about 8000 BC, probably having a symbolic use, were discovered. To the same astonishingly remote period of 8000 BC circa belongs also what was considered up to the 1990s the first stone structure ever: the Tower of Jericho. It is a 8.5 m-tall stone structure which is actually part of an entire circuit of walls discovered at Jericho in the first half of the twentieth century. At the time of discovery these structures were (understandably) attributed to the town described in the Bible and therefore dated to the middle Bronze Age (around 1500 BC). It was only later, at the beginning of the 1960s, that the attribution to a much more remote period was firmly established. The town of Jericho precedes the "revolution" and as such it is a structure which in a sense should not exist, or rather, should not exist if the fallacious idea of the linear evolution of civilisation is strictly adhered to. One of the consequences of such a belief, which—I am afraid—we shall never manage to shrug off completely, is that poor Jericho, along with the slightly later-dated town of Çatalhöyük in Anatolia, is usually categorised as a *proto*-city.

Be that as it may, the tower of such a proto-city is a rather refined building fitted with an internal staircase of twenty two steps, with walls approximately 1.5 m thick. It belongs, as mentioned, to the circuit wall of the settlement. The wall might have been planned for defensive or anti-flood reasons. No suggestion of warfare, however, has been found at the site; in any case a specific defensive purpose for the tower itself looks doubtful. A symbolic explanation, as we shall see, would make more sense.

If Jericho is dated so astonishingly early, the actual date of the beginning of monumental architecture is, incredible as it may seem, older: the tenth millennium BC. This occurred in Gobekli Tepe, a remote hill in the province of Urfa in south-eastern Turkey (Schmidt 2001, 2006, 2010; Dietrich et al. 2013). Here, some 12,000 years ago, elaborate architectural projects were devised, with the construction of stone enclosures endowed with huge, T-shaped megaliths. Unfortunately, up to now only a few of these monuments, whose dramatic importance and longevity were recognized in the 1990s, have been excavated (geophysical prospection shows the presence of many others, giving a total of at least thirty) (Fig. 4.11).

The project of each enclosure at Gobekli is based on a circular or oval wall precinct in which several T-shaped monolithic pillars, usually finely engraved, are nested. Two further pillars stand in the centre of each enclosure, parallel to each other and fixed in sockets carved into the bedrock. Similar, but smaller, rectangular enclosures are also present, relating to a later stage. The monoliths, each weighing several tons, were extracted from a quarry on the hill, where many, unfinished, still remain. As mentioned, materials associated with the structures of Gobekli Tepe yielded amazing dating results: indeed the first buildings date back to a period known as Pre-Pottery Neolithic A (PPNA): the tenth millennium BC. The site was apparently frequented for a couple of millennia, and then backfilled with earth, stones and rubble before being abandoned.

Fig. 4.11 Gobekli Tepe. View of the enclosures from the south-east

The actual discovery of such complex architecture dating back to such a remote period of time is in itself astounding, but this is only the beginning.

Most of the Gobekli pillars are embellished with artistic masterpieces. These include geometric-shaped symbols, and abstract, or at any rate highly stylised, representations of humans. In particular the central pillars appear to represent anthropomorphic beings, with arms and suggestions of clothing, like loincloths. But the vast majority of the art at Gobekli is naturalistic. The reliefs depict animals, represented as single units or engaged in relatively complex scenes, and include mammals (lions, bulls, boars, foxes, gazelles), snakes and arthropods. A special role is played also by birds, several species of which appear. Moreover, figurative art at Gobekli was not limited to carvings, as several standing statues (especially of bears) as well as statues of animals rising in 3D, as it were, from the flanks of the pillars have been found. In the second stage of Gobekli, circular enclosures gave way to rectangular ones, but the figurative art remains; for instance a pair of central pillars exhibits two magnificent carvings of lions, though with more abstract and oblong traits.

Gobekli Tepe is amazing by any standards. It is a complex place, clearly conceived of as a sacred destination and frequented by people from a wide geographical area, well before the "invention" of sedentarism and agriculture. It is obvious that some sort of centralised power, or at the very least a religious influence, must have prevailed at the site, and it is patently obvious that the societies who were able to construct these first temples of humanity were also capable of investing a considerable number of man-hours in a purely idealistic, non-utilitarian project.

As mentioned above, for reasons that have never become clear, this sacred place was meticulously buried, filled in so efficiently as to preserve it as a time-capsule for posterity. In spite of its almost miraculous state of preservation, however, it is difficult to gauge the significance of Gobekli Tepe. The place might have been associated with the dead and the cult of the ancestors, as the anthropomorphic pillars suggest; however, no graves have as yet been found. The presence of naturalistic depictions calls to mind cave art, but dangerous non-mammal beasts (like scorpions and snakes) are unique to Gobekli. What looks certain is that Gobekli was a centre of attraction and pilgrimage; it has even been suggested that domesticated grain might have been first obtained from wild wheat originating in Karaca Dag, which is close by. So perhaps one day we shall really have the proof that it was monumental architecture that triggered sedentarism, and not vice versa.

4.4 The Birth of Astronomically Anchored Monumental Architecture

Having thus reconstructed what we know about the origin of monumental architecture (without ruling out possible dramatic future discoveries that could push it even further back into the mists of antiquity) let us turn to the problem of the birth of *astronomically anchored* architecture, that is—when did people start to incorporate astronomical knowledge into their buildings? Well, the answer seems to be that architecture was linked to astronomy at such an early time that the history of architecture and history of astronomy are virtually inseparable.

Let us proceed along the same path we followed in the previous section. We started with the idea that the megalithic architecture of Europe was the most ancient in the world. The presence of astronomy is undeniable in such places; for instance at Stonehenge and Newgrange (as we shall see in detail in Chap. 7). Going back in time to the middle of the fourth millennium we encounter the temples of Malta. Among them, one can be singled out immediately as being astronomically related: Mnajdra Temple 2. The axis here runs straight east-west up to the west wall, where a central altar is to be found. To the left and right there are two similar monolithic slabs, and the upper external corners of the two slabs define two alignments which pass through the access to the temple and point north of east and south of east in such a way that the window which they span corresponds to the path of the sun on the horizon throughout the year. Thus, Mnajdra works as a solar calendar: the sun rising at the winter solstice creates a flag-like (or axe-like) figure on the right slab (Albrecht 2001). Over the course of the seasons, one can follow the movement of the sun, which rises at the horizon, observing day by day at which point the light strikes the altar in the recess of the temple. In particular, the equinoctial sun rises in alignment with the axis, while when the axe figure is re-formed on the opposite slab, the sun has reached its midsummer rising point. So, astronomy was clearly in operation in Malta; the orientation of almost all other temples is also most likely astronomical, as it is restricted to a region of the horizon which lies between the winter solstice sunrise and due south, and there is the distinct possibility that in some cases—for instance in Ggantija—the builders followed the stars of the Cross-Centaurus constructing a new temple when, due to precession, the previous one got misaligned with them (Foderò Serio et al. 1992).

Let us now proceed further back in time. Even older than the Maltese temples are some of the megalithic structures of northern Europe: the passage graves of Neolithic Brittany, for example, Kerkado. Again, such structures exhibit a clear use of orientation, as *all* of the 68 measurable passage graves face the sun as it rises or climbs in the sky (Hoskin 2007). Equally clear is deliberate orientation in the stone ovals of southern Portugal, like Almendres. There are, in fact, twelve surviving monuments and all are orientated very close to due east (Pimenta and Tirapicos 2008).

Moving further back in time: we have already touched on the enclosures of central Europe, like Gosek. The Gosek circle has three openings or gates, aligned with the sunrise/sunset of the sun at midwinter and with the north direction. Much ink has been spilt in hailing Gosek as "the first observatory"; such claims are extremely dubious, but the fact remains: that the builders of Gosek—and apparently also of several other enclosures—were interested in astronomically relevant directions (astronomical correlation has been claimed also for Warren Field, where it has been suggested that the alignment of pits was deliberately oriented to sunrise at the winter solstice; Gaffney et al. 2013).

Winding back the clock even further, we come across the Neolithic town of Jericho. Recently, a symbolic, astronomically-related function has been proposed for the Jericho tower (Barkai and Liran 2008). Indeed, it has been noticed that, at sunset during the summer solstice, the tower is the first element of the town to be touched by the shadow of nearby mountains; the internal staircase actually gives access to a commanding view of the most prominent peak of the mountain ridge. It has therefore been suggested that the monument was used as an astronomically-related symbol, equipped with a time-keeping function, but also meant to create cultural identity, in line with a mechanism connecting the sky with monumental architecture (which we shall discuss in general terms in next chapter).

So, what we need to conclude our parallel list of the birth of architecture and the birth of astronomy is the place where, as far as we know, monumental architecture actually came into being: Gobekli Tepe. For the sake of completeness, I shall briefly discuss the possible role of astronomy in this place. I stress, however, that this interpretation of the last piece of the puzzle is only very tentative: it is in fact an astronomical reading which I have personally put forward knowing the orientation of only a few enclosures, for the simple reason that all the others still await excavation (Magli 2015; Ananthaswamy 2013). In my defence I can only say that it may take centuries to complete the picture, given the present speed of the ongoing archaeological digs.

The azimuths of the central axis of the three oval enclosures excavated so far (very similar in conception and probably built successively to each other) are all close to due south, but show a slight *decrease*. Simulating the sky in the tenth millennium BC, it is possible to see that a quite spectacular phenomenon occurred at the latitude of Gobekli Tepe at that time: the "birth" of a new star, which started to be visible very low on the southern horizon and had heliacal rising close to the summer solstice. What is striking is that it was the brightest star of the sky, Sirius, that was placed below the horizon by precession in the years around 15,000 BC. After reaching the minimum, Sirius started to approach the horizon and it became visible again towards 9300 BC. Of course extinction effects lowered its magnitude by a significant amount (although much less compared to what it would be today, after the industrial revolution) but the appearance of a new star which became brighter and brighter and culminated higher and higher over the course of the centuries would have been a noticeable phenomenon, and this was apparently followed over the centuries by decreasing the azimuths of the axes of the enclosures (Magli 2015) (Fig. 4.12).

The image of the "birth" of Sirius might also be present in the iconography at the site—to be exact, on one of the most elaborate of the pillars, Megalith 43 of structure D. Curiously enough, this pillar has many similarities to the (of course much later) Babylonian boundary stelae, called *kudurru*, which contained detailed references to the sky. In particular, these stelae contained a register (one of the bands into which the stelae are divided) with "box" altars devoted to the gods of the sky which are very similar to the "bags" appearing in the upper register of the Gobekli pillar. Several constellations were depicted in the lowermost register of the Kudurrus (Iwaniszewski 2004), again in a way similar to the Gobekli pillar if

Fig. 4.12 Gobekli Tepe. 10 Millennium BC. Sirius starts to be visible

the figures there—a scorpion, a goose, a fox and a beheaded ithyphallic man—are taken to be constellations. Besides this analogy, what is really impressive is the scene in the middle register of the Pillar. In front of a rectilinear band of small squares and v-shaped motifs we find a vulture who seems to be raising a circular object and two wader-birds, the lower one perhaps being newborn (a serpent or phallic symbol, and two bone-shaped or H-shaped carvings complete the scene). There is no doubt that the vulture has human-like features: the expression resembles a smile and the way it stretches its wings is unnatural for a bird but would be very natural for a man wearing fake wings. So it is likely that the image depicts a therianthrope (a hybrid human-animal creature) or, more simply, a shaman wearing a vulture costume. Vultures are connected with the dead and appear together with decapitated human bodies in the art of Catal Huyuk, which is chronologically not so remote from Gobekli. However, the vulture has also been repeatedly associated with the sun in many cultures (an obvious case being ancient Egypt). For this reason, I proposed that the scene might represent the "birth" of Sirius, borne aloft by the sun. The day may be the summer solstice, since in the centuries around 9300 BC, the summer solstice was on the point of leaving Scorpio and was therefore just above the head of the animal. Of course Scorpio was invisible at the solstice, being just below the horizon, under the sun: in a sense Scorpio was in the realm of the dead, as indeed it is on the stela, along with a decapitated man, in the register beneath the main scene (Figs. 4.13 and 4.14).

It goes without saying that future discoveries at Gobekli could modify or even invalidate this interpretation, which therefore must remain in the realms of speculation for the time being. What is certain, though, is that sooner or later we are due to be dazzled when some thrilling new information comes to light. In fact, Gobekli is not unique. There are other centres located in the outskirts of the Urfa plain which

Fig. 4.13 Gobekli Tepe.
Pillar 48

Fig. 4.14 A Boundary stone
(kudurru) from southern Iraq,
twelfth century BC. The text
records the granting of lands
to a man called Gula-eresh.
Nine Mesopotamian gods are
invoked to protect the
monument, with a series of
celestial references. In the
upper register the solar disc
(the sun-god Shamash), the
moon crescent (the Goddess
Sin) and Venus (Ishtar). The
boxes beneath represent altars.
In the lower register appear 3
constellations, Draco, Lio and
Scorpio. *Courtesy* © Trustees
of the British Museum

are certainly contemporary with Gobekli, all furnished with T-shaped monolithic pillars whose heads barely emerge from the unexcavated terrain (Celik 2001; Güler et al. 2012).

To conclude then, our poor non-sedentarist, pre-revolutionary people who lived in what we clinically call the "Pre-Pottery Neolithic A" were blissfully unaware that they were living in a period that was later merely classified as *"pre-"* something.

Even so, they have only just started to reveal their secrets to the smug sedentarists of today.

References

Albrecht, K. (2001). *Maltas tempel: zwischen religion und astronomie*. Potsdam: Naether-Verlag.

Ananthaswamy, A. (2013). Stone Age temple tracked the dog star. *New Scientist, 219*, p14.

Barkai, R., & Liran, R. (2008). Midsummer sunset at Neolithic Jericho. *Time and Mind: The Journal of Archaeology, Consciousness and Culture, 1*, 273–284.

Cauvin, J. (2000). *The birth of the Gods and the origins of agriculture*. Cambridge: Cambridge University Press.

Celik, B. (2001). Karahan Tepe: a new cultural centre in the Urfa area in Turkey. *Documenta Praehistorica, XXXVIII*.

D'Errico, F. (1989). Palaeolithic lunar calendars: A case of wishful thinking? *Current Anthropology, 30*(1), 117–118.

D'Errico, F., & Cacho, C. (1994). Notation versus decoration in the upper palaeolithic: A case-study from Tossal de la Rocca. *Journal of Archaeological Science, 21*, 185–200.

Dietrich, O., Köksal-Schmidt, Ç., Notroff, J., & Schmidt, Klaus. (2013). Establishing a radiocarbon sequence for Göbekli Tepe. *State of Research and New Data Neo-Lithics, 1* (2013), 36–47.

Foderò Serio, G., Hoskin, M., & Ventura, F. (1992). The orientations of the temples of Malta. *Journal of the History of Astronomy, 23*, 107–119.

Gimbutas, M. (2001). *The language of the goddess*. London: Thames & Hudson.

Güler, M., Celik, B., & Güler, G. (2012). New pre-pottery neolithic settlements from Viransehir district. *Anatolia, 38*.

Hoskin, M. (2007). Orientations of neolithic monuments of Brittany: (2) The early dolmens. *Journal for the History of Astronomy, 38*, 487–492.

Iwaniszewski, S. (2004). Archaeoastronomical analysis of Assyrian and Babylonian Monuments: Methodological issues. *Journal History of Astronomy, 34*, 114.

Lawson, A. J. (2012). *Painted caves: Palaeolithic Rock Art in Western Europe*. Oxford: Oxford University Press.

Lewis-Williams, D., & Pearce, D. (2005). *Inside the Neolithic mind: Consciousness, cosmos and the realm of the gods London*. London: Thames & Hudson.

Lull, J. (2014). La alineación solar del equinoccio en la Cova del Parpalló: una nueva aproximación arqueoastronómica. *Huygens, 107*, 4–10.

Mithen, S. (1998). *The prehistory of the mind: A search for the origins of art, religion and science*. New York, NY: W&N.

Magli, G. (2015). Sirius and the project of the megalithic enclosures at Göbekli Tepe. *Nexus Network Journal, 17*, 1–11.

Marshack, A. (1972). *The roots of civilization*. London: Weidenfield.

Pásztor, E. (2014). Prehistoric astronomers? Ancient knowledge created by modern myth. *Journal of Cosmology, 14* (2011).

Pimenta, F., & Tirapicos, L. (2008). The orientations of central Alentejo megalithic enclosures. In J. Vaiskunas (Ed.), *Astronomy and cosmology in folk traditions and cultural heritage. Archaeologia Baltica* (Vol. 10, pp. 234–240). Klaipeda: Klaipeda University.

Rappenglück, M. (1998). *Palaeolithic shamanistic cosmography.*

Rappenglück, M. (2008). Astronomische Ikonographie im Jungeren Palaolithikum. *Acta Praehistorica et Archaeologica, 40,* 179–203.

Rappenglück, M. (2011). Earlier prehistory. In C. L. N. Ruggles & M. Cotte (Eds.), *Heritage sites of astronomy and archaeoastronomy. ICOMOS–IAU* (pp. 13–27), Paris.

Rappenglück, M. (2012). Stone age people controlling time and space: Evidences for measuring instruments and methods in earlier prehistory and the roots of mathematics, astronomy, and metrology. In *5th International Conference of the European Society for the History of Science.* Athens.

Runciman, W. G. (2001). The origin of human social institutions. In *Proceedings of the British academy* (Vol. 110). Oxford: Oxford University Press.

Schmidt, K. (2001). Göbekli Tepe, Southeastern Turkey. A preliminary report on the 1995–1999 excavations. *Paléorient, 26*(1), 45–54.

Schmidt, K. (2006). Sie bauten die ersten Tempel. Das rätselhafte Heiligtum der Steinzeitjäger. Die archäologische Entdeckung am Göbekli Tepe. München: C.H. Beck.

Schmidt, K. (2010). Göbekli Tepe: the stone age sanctuaries. New result of ongoing excavations with a special focus on sculptures and high reliefs. In M. Budja (Ed.), *17th Neolithic Studies. Documenta Praehistorica, 17,* 239–256.

Sparrow, T., McMillan, A., Cowley, D., Fraser, S., Murray, C., Murray, H., et al. (2013). Time and a place: A luni-solar 'time-reckoner' from 8th millennium BC Scotland. *Internet Archaeology, 34,* 1.

Trump. (2002). *Malta: prehistory and temples.* Midsea Books.

Chapter 5
Astronomy, Power, and Landscapes of Power

5.1 Sky and Cosmos

The connection between astronomy and daily life in ancient times—at least after the definitive establishment of sedentarian life—was far more complex than a simple understanding of natural cycles. We can be sure of this because the sky was linked to a fundamental mechanism of social dynamics: the management of power. To grasp the connection between astronomy and power, however, first we must make a detour to understand the concept of *worldview*.

Human beings need a worldview, a *Weltanschauung*. Our worldview includes a scientific explanation for a series of things which at first glance appear mysterious and inexplicable: earthquakes, eclipses, viral epidemics and so on. In spite of this, most of humanity is of religious inclination, and this means that our worldview also encompasses the belief in "another" world. In ancient times, religion was the only available key to the inexplicable, and above all, the key to the mystery of human illness and death, the incomprehensible and the deep-seated refusal to accept the end of life. As a consequence of the centrality of this notion, religion also acted as a key to temporal power. Accordingly, as we shall see, the monuments which are the object of study of archaeoastronomy were built to serve as witnesses to power. The connecting thread that underpins this, which we shall now study in some depth, is, however, very simple and can be put in a nutshell.

From nature, a worldview is built, and from this worldview, power is legitimised; monumental architecture is then created to act as explicit witness to the legitimacy of power. As a consequence, it has to be explicitly related to the natural cycles which regulate the worldview. It follows, thus, that it usually incorporates a series of symbolic references to the sky.

To understand how a worldview is formed, we start from a simple fact. Human life happens to be located on earth, but humans do not accept the earth as the only place where their whole sphere of existence, everything relating to their own life, is located. They prefer to think, rather, that their place is in the Cosmos. The Cosmos

© Springer Nature Switzerland AG 2020
G. Magli, *Archaeoastronomy*, Undergraduate Lecture Notes in Physics,
https://doi.org/10.1007/978-3-030-45147-9_5

is a space-time entity, since it is a place—a cosmic, or sacred, place—where life is regulated according to an appropriate, cosmic time. Both cosmic space and cosmic time are related to the sky.

Cosmic time derives from the observation of the celestial cycles. As a matter of fact, although there exist extraordinary phenomena in the sky which are difficult or impossible to predict such as eclipses or meteors, most celestial phenomena are eminently predictable: it is sufficient to study them for a suitable amount of time and with suitable precision. This is the only way to interact with the sky: forecasting. Forecasting an event confers power on the one predicting it and affords comfort to the one awaiting it—for example the heliacal rising of a star which returns in the sky at the predicted time.

Scheduling events in a reassuring cyclical succession thus puts order into time. Time ordered in this way then becomes sacred, as it has been properly inserted into the worldview. The natural cycles by which cosmic time is marked are naturally divided into two: diurnal and nocturnal. In the diurnal sky, the sun marks the rhythm of the seasons and thus governs the swings between hot and cold, rainy and dry seasons, sowing and harvesting. The cycles of the Earth and that of the Sun are indeed connected. By following the movement of the sun on the horizon day after day, one soon realises that our star rises more and more to the south of east when approaching the winter solstice. This might even prompt the idea that the sun is heading south irrevocably and doomed not to rise again, but—reassuringly—the velocity with which the rising sun goes south slows down to zero at the solstice; for a few days the difference in the rising point (which is nonetheless present, since the extremum is reached only on the day of the solstice) is too small to be appreciable, until finally the sun is seen moving again towards the east. This process is naturally correlated with a similar seasonal process of dying and rebirth of the natural cycles, the amount of daylight throughout the year, health, and so on. As a consequence, the winter solstice was viewed as a harbinger of rebirth, possibly even of a successful earth-sky re-fertilisation, and the timing of the solstices was in many cases the fundamental marker of cosmic time.

If, on one hand, keeping track of the sun's behaviour is basic to human activity, on the other, the man who understands and predicts the solar cycle becomes an intermediary, the guarantor of the annual repetition of the cycle. In turn, the night sky has two principal bright objects: the moon, whose movement over a month seems to imitate the sun's over a year; and Venus, whose behaviour alternates between announcing the sun's rising and following the sun's setting. Furthermore, there is the regular, perennial motion of the stars. It is during the night that the gates to the afterworld are ideally opened, and so yet again, the man who controls nocturnal cycles can credit himself with being an intermediary with the sky.

Cosmic space is closely connected with the sky. In fact, cosmic space is an ordered—typically quadripartite—place, and the rule on which it is founded is based on the celestial cycles. The first main axis, the north-south one, comes naturally from the observation of the night sky. Accurate observation of the star's turnaround leads us in fact to conclude that there exists a fixed point, a centre of the heavens (which may or may not coincide with a star, depending on the epoch), the celestial pole. This point allows us to define a privileged axis, an *axis mundi*, as it is sometimes defined, as the direction from the observer to the pole. Finally, this direction can be projected on the ground to find the meridian, which is thus associated with the stars, and consequently with the night. The orthogonal direction, east-west, is instead usually associated with the sun and hence with the daylight. It can be found by taking the perpendicular to the meridian, but it can also be found independently by observing the sun's movement during the day, and connecting his rising/setting point (if the horizon is flat). Analogously, the south (in the northern hemisphere) can be found independently from nocturnal observation by bisecting the rising/setting direction of the sun, measured on a circle, or by following the shadow of a stick around noon.

The result of these operations is a division of the world into four parts, which is both symbolic and practical. It is a "cosmisation" procedure, in which space becomes ordered, a place which has been founded and prepared for human life. The historian of religion Eliade (1959) was probably the first to understand these mechanisms in details. He wrote:

> The Experience of Sacred Space makes possible the founding of the world: where the sacred manifests itself in space, the real unveils itself, the world comes into existence. To be inhabited, the world must first be founded.

Since the world is divided into four parts, there is a point at which the axes intersect and the four parts become one. It is a privileged point, a centre, the navel of the world. Each culture had its own cosmos, and therefore each culture had its own umbilicus: all Roman roads led to Rome, all Inca roads led to Cusco, while for Dante Alighieri, Jerusalem was at the centre of all lands.

The navel of the world is naturally a special place, the ideal place for communicating with other levels, the place where power resides and manifests its identity physically. So, it is often also a place where the cosmos is represented, even replicated, as well as the elected place for embarking on the afterlife. In some tombs of Maya rulers, for example, the walls of the chamber are frescoed with glyphs of the cardinal points, and similarly, the Egyptian pyramids were precisely oriented to the cardinal points, and the Egyptian tombs had "magic bricks" placed in the walls, one for each of the cardinal points.

5.2 Cosmos and Afterworld

The afterlife is a structural component of any worldview, since the cosmos must accommodate a solution for the afterlife. As a consequence, the cosmos is "tiered", organized into levels. Of course one then needs gates, allowing transition between levels. During the Palaeolithic period, for instance, natural caves may have assumed the significance of symbolic places of rebirth, and consequently places of power. With the birth of monumental architecture, monumental replicas of caves began to be built, in the form of chambered cairns and tombs. In a sense, however, these places—as transitional, and therefore potentially dangerous, places—have to be isolated, as indeed the caves are. This may have been achieved by the use of water, filling moats encircling the graves, or even with specific choices of location in the landscape, such as the bend of the river Boinne (Sect. 7.2), which naturally offers a sort of promontory where Newgrange and other Neolithic tombs were to be built.

With the birth of monumental architecture, then, the sky and the earth, the underworld and the cosmos meet in specially prepared places. The word "tomb" may be reductive here: monumental graves like Newgrange were probably also sites of pilgrimage, and many of the temples associated with the great pyramids of Egypt functioned as shrines of worship of the corresponding divinised Pharaohs for many centuries. What must always be remembered is that burying the dead, worshipping ancestors and the like were actions *directed towards the living*. Places built to act as an interconnection between levels actually became physical intermediaries between the sacred space and the sacred itself, so that the pyramids of Egypt—for instance—were artefacts aimed at assuring the afterlife of a ruler and *consequently* the correct and predictable progression of the life of the living thereafter.

As a good example of the simultaneous operation of all these mechanisms we might cite the so-called Temple of the Inscriptions in the Maya site of Palenque, Chiapas (Fig. 5.1). The monument was built during the reign of King Pacal (603–683 AD) and is one of the masterpieces of Mayan architecture, a nine-stepped pyramid with a small structure on top (Schele and Freidel 1990). The Maya afterworld or Xibalbà was tiered into a complex geography of nine levels, a clue that the temple probably had something to do with this. This was not known, however, until 1952, when the Mexican archaeologist Alberto Ruiz discovered that

Fig. 5.1 Palenque. The temple of the inscriptions

Fig. 5.2 Palenque. The lid of Pacal's sarcophagus. Usually depicted with the longest side set horizontally, it is instead easier to understand if viewed vertically as here, actually the same view allowed by the entrance to the tomb

the temple was also the tomb of the king, who was buried in a chamber located at the base of the monument and accessed through a staircase from the top. The tomb, which had been constructed before the erection of the pyramid and was therefore part of a single architectural design, was intact. Pacal was resting in a stone sarcophagus, sealed with an elaborately inscribed lid. The image depicted on the lid, possibly the most famous Mayan work of art, shows the king amid a scene that bursts with complex astronomical-cosmological significance (Fig. 5.2). Pacal is depicted at the centre of the scene, which is traversed by a cross-shaped figure probably representing the intersection between the Milky Way and a band symbolising the Ecliptic. Beneath him there is the entrance to Xibalbà, the Kingdom of the Dead, guarded by a monster-like mask and represented materially by the nine tiers of the building constructed above the tomb. But the death of the King must also be a rebirth, since his son has to reign after him and his people have to live on. To understand how Pacal is associated with rebirth we need to make the following observation. The two zones in which the Milky Way crosses the Ecliptic mark, naturally, the rising of the sun on two days of the year. Due to precession, these two dates vary over the course of the centuries, but during the classical Maya period these dates were not far from the solstices. Hence Pacal is ideally placed at the solstitial point, so that his death might be identified with rebirth, in analogy with the resumption of the solar cycle and thus of agricultural activities, particularly the cultivation of maize (as an explicit reminder of the connection between the deceased and the rebirth of the solar cycle, close to the days of the winter solstice— looking from the plaza and the palace in front of the building, the sun sets behind the Temple of Inscriptions, ideally descending into the crypt). Interestingly, although the tomb was completely inaccessible after the king's interment (the descending staircase was filled with earth and rubbish) a sort of magical liaising with the world of the living and the world of the spirits was nevertheless assured by means of a small shaft which runs parallel to the staircase and was supposed to allow Pacal's spirit to emerge and manifest itself. This intriguing image is represented explicitly in a panel of one of the temples built by Pacal's son, the so-called Group of the Cross. The image makes it clear that communication with the dead occurs in an altered state of consciousness, triggered by the use of drugs, and this leads us to another fundamental point, namely, the ways of mastering the Cosmos.

5.3 Mastering the Cosmos

Individuating the Cosmos—directions, celestial cycles, axis mundi, and the like— and opening up communication between cosmic levels are thus fundamental operations for human societies. The performance of such operations, consequently, yields power (Krupp 1997).

Keeping track of the annual solar cycle links power with the calendar and the regular succession of the seasons. Monitoring the cycles of the night stars links power with the regular development of the hours of the night in expectation of the return of the sun. Establishing the cardinal axes and their centre links power with the very heart of human existence. Managing the calendar for the festivities of the Gods and the worship of the ancestors links power with religion and the afterlife. And so on.

We must now study very carefully the identity and the role of the people who mastered such powers since, of course, these same people were also the originators of the projects of sky-oriented power architecture. Perhaps we should call them priest-astronomers. Here, we shall use the traditional term *shamans*.

The historian of religion Mircea Eliade defined shamanism as a technique for generating religious ecstasy (Eliade 1964). A Shaman is in fact a person who claims to be a mediator between the human and the supernatural worlds. The shaman can gather information which is forbidden to common people, and is able to have contact with the "other world" as well as to convey to the world of the spirits requests from the humans such as, for instance, the desire to recover from illnesses or for rains to return after a drought. It is important to stress at this point that there is no need for the reader to believe in the existence of any special shamanistic power (as far as the author is concerned, as a physicist I do not believe in any "paranormal" powers whatsoever). What matters here is only what shamans *claim* about themselves and what people are prepared to believe about them.

Shamans claim to reach their higher levels of communication through experiencing altered states of consciousness, typically induced by drugs. To understand these practices, ethnological-anthropological documentation regarding present day people's experiences is invaluable. In fact, practices very similar to present day ones are archaeologically and historically documented in numerous parts of the world. For example, shamanistic practices involving hallucinogenic-induced states of ecstasy were practised in Mesoamerica, and there is overwhelming evidence to support the fact that the religion of the Upper Palaeolithic-Neolithic period was also a form of shamanism. The latter idea is mainly based on the images which appear in chambered tombs (see Sect. 7.2). These include geometric carvings—mostly spirals and ovals—which are indistinguishable from those inspired by altered states of consciousness in modern-day populations who make use of hallucinogenic substances, and from depictions produced by volunteers in experimental tests (Dronfield 1995) (Fig. 5.3). Moreover, in cave art there are representations of therianthropes (half man-half animal beings), and these recur also in modern day descriptions of visions induced by hallucinogens.

Let us analyse then what is known about shamanistic practices. The trance state allows the Shaman to "open the gates", and even—it is thought—to leave his temporal body and to travel in the supernatural world. This travelling usually occurs through the sky, so that birds and men-birds feature prominently in the shamanistic iconography. In his cosmic journeys the shaman uses maps which, at least in

Fig. 5.3 Newgrange. Spirals and other "abstract" carvings on kerbstone 52

principle, encompass the whole universe. In particular, the link with the axis of the world is central to the shaman's iconography, so shamans usually carry a wooden stick. Other status-symbols objects or paraphernalia are also common, for example, the drum. The drum is associated with musical rhythm, and the generation of obsessive, cadenced melodies helps in creating a suitable atmosphere for altered states of consciousness.

As an example of shamanistic rites, consider the Altaic populations of central Asia (Eliade 1964, 1971). For these people the sky is the kingdom of a god, called Bai-Ulgan, and it is divided into nine levels, each of which is assigned to a sister of the god. This celestial kingdom can be visited by the shaman via a mystic flight in trance state, during which the shaman imitates the duck's call. A birch pole carved with nine notches is placed at the centre of the shaman's tent and represents the central axis of the world, the ideal extension of the polar axis. So, the shaman's tent is an idealised replica of the heavenly vault, and the point at which the axis passes into the earth is the centre of the world: in a sense, this provides certainty that the world exists and is thus able to be inhabited: the Altaic people can be certain of living "in the right place".

As intermediary between the human and the other levels, the shaman is the elected custodian of celestial matters. He is therefore usually also the custodian of the calendar. He directs the rhythms of rites and celebrations, and is in a sense responsible for the cyclical renewal of nature. Consequently, his role usually implies knowledge of a great deal of notions, a grounding in natural sciences and a sprinkling of traditional lore, all of which the shaman acquires during his apprenticeship with a senior shaman and which help him in carrying out his duties. For example, the shaman acts as healer with the sapient use of medicinal herbs and plants. Furthermore, the ability of the Shaman to communicate with the supernatural, in many cases linked with his knowledge of astronomy, reveals another aspect of his powers: divination. The production of omens is thus another typical duty—and power—of priest-astronomers. For example, in China there exist records of astronomical divination that can be traced back to the Shang dynasty (1766–1123 BC). These texts show that divination was practised at the imperial court and that it was deemed to be a matter of state, capable of influencing decisions at the highest level. Similarly, court astrologers/astronomers capable of extremely precise observations were fundamental in Mesopotamia (Steele 2007). Among the Maya, the surviving codices tells us that astronomy and astrology were deeply interlinked and that, in particular, a special role with regard to warfare was played by omens connected with the cycle of Venus. The importance of omens can be traced on up to Greek and Roman times, especially in the form of horoscopes, to the point that many Roman emperors had personal astrologers—according to historians—and based their decisions on omens indicating whether a day would be auspicious or not.

This finally leads us to a delicate matter, that of understanding the relationship between shamanism and temporal power. This problem appears to have been somewhat overlooked in anthropological literature, but for our purposes the key elements are clear. Power must be conferred by someone else. In a democracy, power is (in principle at least) conferred by the people, but with a kingship the only

Fig. 5.4 Edfu. The Pharaoh and the Goddess of knowledge Seshat put in place together the foundations of the Edfu temple. The texts says that they are looking at the northern stars

way to receive it is by investiture from above. Therefore, the foundation of power and the Cosmos are inextricably connected. Since the shaman is the intermediary with the cosmos, kings and shamans are likewise connected. The power of the shaman and temporal power are usually distinguished by their very nature, since shamanistic power is transcendent, and only very rarely did temporal power succeeded in re-uniting both aspects in a single figure. One striking exception is to be found in ancient Egypt, in the cosmovision devised for/by the Pharaohs: here the king is identified with a living god, automatically conferring shamanistic powers on the king himself as an intermediary with the gods (Fig. 5.4).

5.4 Cosmic Machines

The underlying ideology behind shamanistic power and religion is inevitably and invariably the same: power asserts itself by forging a bond with nature. The need then arises to exploit such a bond and make it tangible. This is in part attained with transitory, occasional events. For instance, special one-off public ceremonies (like the coronation of a new king or the funeral of a dead one) and recurrent public rituals and feasts, connected to the calendar. It is also effected by the creation of shared, easily recognisable symbols or images, like banners, insignia, coats of arms, uniforms and also with oral or written traditions and myths. But this is not enough.

Establishing cultural—or perhaps I should say corporate—memory, in which the established (and establishment's) worldview is accommodated also requires the cosmovision to be made tangible, immanent, in a perennial (or purportedly so) form. Monumental architecture thus stems from—and is a symbol of—power, and has to be explicitly connected with the cycles of the cosmos. It is here that astronomy comes into play, through a mechanism which had previously been understood in broad terms within the history of religion, independently from the development of archaeoastronomy: the *hierophany* (Eliade 1964).

A hierophany (from a combination of Greek words, meaning "the sacred reveals itself") is an explicit manifestation of the sacred: of a god, of a message from a god, or also of a divine aspect of nature. Simple examples of hierophanies are just the various means of communication with divine power. For instance, in official festivals the power of the gods is symbolically renewed, and in cosmological myths a fusion between human and divine (an irruption of the sacred into the immanent, Eliade would say) is usually present. The hierophanies of myths have a pedagogical content, since the manifestation of the sacred is used for the construction of models which have to be imitated, or even of models which power is seen to imitate. These models can be approximated or "returned to" through religious life and collective religious behaviour. The concept of hierophany is not, however, exhausted by these examples. In fact, the sharp distinction between the sacred and the profane worlds, which is slightly blurred in these examples, can be *dramatically* shattered by a far more spectacular kind of event. To understand this, notice that the distinction between the sacred and the profane was perceived also at the level of physical space and of the human environment. As we have seen, space is validated and has to *deserve* to be lived in, only if it conforms to patterns and rules determined by the sacred, such as orientation. In a sense then, the only true space is the sacred space. But if the space is sacred, then it is also an arena where the divine can—or even must—manifest itself. So, a hierophany can be a truly physical, tangible manifestation of the sacred in the place that is worthy of allowing such a manifestation to occur. Eliade wrote:

> When the sacred manifests itself in any hierophany, there is not only a break in the homogeneity of space; there is also a revelation of an absolute reality, opposed to the non reality of the vast surrounding expanse…the hierophany reveals an absolute fixed point, a center.

Hierophanies can therefore appear as abrupt flashes occurring in sacred spaces: architecture is the means by which they are brought to the highest level of emotional participation, and astronomy is the tool for engineering such events. An architectural hierophany can therefore be described as a mechanism whereby appropriately designed elements combine with celestial elements. Consider, for example, a building aligned with the sun that rises or sets in a particular direction, and consider it together with the sun itself in its constant motion as a single entity, a mechanism made up of different gears. When the gear in motion (the sun) becomes aligned with the given, fixed direction, the mechanism is set in operation. As a result, a special phenomenon occurs; for example, the sunlight penetrating an

Fig. 5.5 Giza, summer solstice. Akhet, the name of the Great Pyramid, is re-written once a year from 4500 years on the Giza plateau, by one of the most complex and spectacular hierophanies ever conceived (see Sect. 8.2 for details)

otherwise dark corridor. If this mechanism has a religious connotation, one can view these phenomena as divine manifestations. In a way, then, despite being objects produced by human endeavour, sacred buildings take on a life of their own at the very moment they become the instruments of such manifestations. As we shall see in greater detail later on, these spectacular events can be divided into three main groups:

- hierophanies directly related to symbols of temporal power with a sacred value (like, for instance, the hierophany occurring in the Pantheon on the day of the mythical foundation of Rome, Sect. 10.4).
- manifestations of the renewal of a god's powers in line with natural cycles (like the alignment to the winter solstice sunrise of Karnak's temple in Egypt, Sect. 8.4, or the Serpent Equinox at Chichen Itza', Sect. 9.2).
- symbols of renewal relating to the afterlife (like the alignment of Newgrange, Sect. 7.2, or the Akhet hierophany at Giza, Sect. 8.2) (Fig. 5.5).

The final category are the most frequently occurring. Many hierophanies were in fact bound up with the only event that power was *not* capable of foreseeing or controlling: death. Thus, those in power chose the message of the hierophany to demonstrate the ability to communicate, control and even survive after death.

In any case, in all such examples, it is the cyclical behaviour of the celestial gears that assures the rigorous, reassuring repetition of the hierophany. The regularity of each cycle is a manifestation of sacred time, and this aspect was also first pointed out by Eliade, who stressed the idea of a "eternal return". Cyclic hierophanies are a recalling of, or even a repetition of, mythical events, which act as sacred promises (for instance of a good harvest) stipulated with the gods. In some cultures the entire year was even seen as a replica of an entire mythical age, and so the entire cosmos ideally underwent a complete cycle, from birth to rebirth, over the course of a year.

An important aspect connected with this comforting cyclical view of the cosmos was the cyclical repetition of pilgrimages and, in many cases, the occurrence of hierophanies as a core feature in them. A pilgrimage is therefore an important social activity linked with monumental architecture and astronomy. From the anthropological point of view, pilgrimages can be interpreted in two, almost antithetical, ways. In Durkheim's (1965) view, religious behaviour is induced by political processes; in other words, they are one of the reflections of the way power manipulates the worldview to legitimise itself. This view appears as the most appropriate in those cases in which pilgrimages were regarded as "state affairs", organised and stage-managed at the highest level. Owing to our lack of information on the subject it is difficult to apply this model to the Neolithic period, but it is very evident in other significant settings, such as the Inca state pilgrimage site of the Island of the Sun (Sect. 9.4).

Another viewpoint with regard to the birth and development of some pilgrimage sites exists, however (Turner 1969). These writers see pilgrimage as a challenge to the establishment—state, or religion, or both—in that people with a "pilgrim status" can more easily evade censuses and slip between social castes. At any rate, a fundamental characteristic of a pilgrimage is its "liminality", its sense of the extreme: a pilgrimage is an arduous endeavour, it involves physical preparation, suffering and privation, it involves travelling in hostile and/or unknown places, and so on. Clearly then, experiencing hierophanies as a culmination of a pilgrimage will render it an even more potent and awe-inspiring experience.

5.5 Cosmic Landscapes

One feature of sacred space is its association with cultural memory. If a place becomes sacred, for whatever reason, the collective memory of the community will tend to preserve this concept of sacredness over time—and hence, typically, the end use of the site and the traditions linked with it. If hierophanies are associated with power at a certain time in a certain place, power will subsequently tend to confirm that very same place as having its *own* natural right to hold sway. All this led to an important development: the fact that some places first acquired and then conserved for centuries or even millennia their reputation as sacred—and therefore powerful—places. As a consequence, each ruler wished to associate his name and his memory with that place, so that the landscape was modelled, modified, reconstructed and re-interpreted with this aim in mind over the years or the centuries. The result is a sacred, *cosmic* landscape: it is a place where the religious element is crucial, and so is made to merge with the natural surroundings, enhanced by human construction.

Fig. 5.6 Balamkamche. The "Ceiba tree" formed by natural concretions in the main ambient of the cave

Usually, specific rules would be established to regulate the development of the landscape in accordance with worldview criteria—in particular, topographical and/ or astronomical rules—and such rules were observed down the centuries.

The reasons behind the selection of such places are often difficult to pinpoint. In some cases, however, it is nature herself who can take responsibility for indicating a place as sacred (Devereux 2010). For instance, the Chumash Indians identified the access to the earth in a natural vagina-shaped formation, the so-called Painted Rock (Krupp 1997). In the Balamkanchè cave of the Yucatan, the Maya discovered a huge subterranean room with, at its centre, a natural column created by the joining of a stalactite and a stalagmite over millennia of water percolation (Fig. 5.6). This natural formation bears an impressive resemblance to a Ceiba, the sacred Maya tree. As a consequence, the place became a shrine, and people brought offerings in the form of ceramic vessels bearing depictions of the mask of the Rain God, which can still be seen in situ as, miraculously, the cave escaped the iconoclastic wrath of the Conquistadores. A further example is the already mentioned Island of the Sun on the Titicaca lake, which the Incas identified as their place of origin. Here, huge natural etchings on the rock resemble giant footprints and were seen by the pilgrims as "footprints of the sun" not far from the sacred rock where the sun itself had allegedly been born (Fig. 5.7).

Similarly, some famous sacred places in the classical world were apparently chosen for their natural characteristics. This holds, for instance, for the most important sanctuary of the classical age, the oracle of Apollo at Delphi. The oracle

Fig. 5.7 Island of the Sun,
Titikaka Lake. The footprints
of the sun

was incarnated by a priestess, the Pythia, who received pilgrims in a cavity located beneath the main temple; here she produced her (notoriously ambiguous) omens in a state of trance. The historian Plutarch writes that the place had been discovered by pure chance and that a sort of perfume could often be detected in the oracle's lair. As a matter of fact, recent research has shown that, under the temple, the friction between two geological faults produces hydrocarbon vapours which can cause a state of euphoria, or even altered states of consciousness (De Boer 2014) (Fig. 5.8).

Finally, this kind of sensitivity to a "naturally sacred" landscape might have played a role in the choice of sacred sites as far back as the Palaeolithic period. For instance, the choice of the Chauvet cave—among dozens present in the area—for the execution of stunning cave paintings masterpieces might have been influenced by the nearby presence of Pont d'Arc, a peculiar rock formation spanning the Ardeche River, 60 m wide and 54 m high. It is a somewhat unique natural feature and, what is more, when seen from a distance it resembles a standing four-legged animal—or by a stretch of imagination, a lion.

The final result of the cosmisation process of an entire landscape is that the human environment is modelled, modified, reconstructed, even created afresh, in such a way that the explicit adherence to rules implies sacredness: human and otherworldly planes communicate through topographical and astronomical alignments. In some cases, all this occurs through *replicas*: the human level reflects the cosmos and becomes a copy of it.

We are touching on a delicate issue here since many of the inane theories available in the (alarmingly vast) pseudo-archaeoastronomy section of pseudo-Archaeology are based on an over-interpretation or misinterpretation of this idea. For instance, it is supposed that the temples in a site, or the pyramids in a Necropolis, or even very distant, non-inter visible cathedrals were arranged on the ground in the same manner as (opportunely selected) stars of some handy constellations. Actually, we have no proof whatsoever that this kind of global "copy of

Fig. 5.8 Delphi. View of the upper terrace of the temple

the sky" was ever attempted and, ironically enough, it is usually *much more complicated* than this. It is not possible to extract a map from the sky and then use it as a handful reference to read the meaning of a site. It was, in fact, the interest in the natural elements *as a whole* that led to the conception of sacred landscapes that would in some way replicate the cosmos. The role of astronomy in such a mechanism was in many cases fundamental *but not the only one* in connecting the various elements, as we shall now illustrate with a few examples, the first being a very special place called the *Heart of Neolithic Orkney*.

The Orkney are a group of small islands located where the North Sea and the Atlantic Ocean meet. The largest, on which we shall focus here, is called Mainland. The islands are fertile and were populated by Neolithic people around 4000 BC. These people developed a complex society whose religious and social heartland was located in western Mainland, which is characterised by two huge sea inlets, the "lochs" of Stenness and Harray, divided by the Ness o' Brodgar, a short strip of land. The main structures in this area have miraculously passed down to us virtually intact, so that the ensemble has been declared a Unesco World Heritage site under the above mentioned name of the Heart of Neolithic Orkney. This sacred centre—which remained in use up to the Bronze Age—was made up of huge ceremonial meeting points, monumental tombs, and villages (most probably for the priests). The main ceremonial places are the Standing Stones o' Stenness and the Ring of Brodgar (Fig. 5.9).

Fig. 5.9 Mainland. The Ring of Brodgar

Stenness was a circle of 12 huge stones (only four remain standing) located by the south-eastern shore of the loch. Excavations have dated the site to about 3100 BC. The monument was a henge, that is, a circular ditch enclosed in a bank, with the ditch excavated directly from the rock surface. It was probably accessed from the north, through an entrance which points towards the contemporary settlement called Barnhouse. In the centre, a stone hearth was placed. The Ring o' Brodgar, strategically located on the Ness between the two lochs, was probably built later. It is a 104 m wide stone ring, of which 27 stones remain. The megaliths are smaller than those of Stenness, but the diameter of the monument is the third largest to be found among the stone circles of the British Isles. The greatest construction effort was put into excavating the rock-cut ditch, which originally had two entrance causeways. The position of the monument is quite dramatic, and it was even more so at the moment of construction, since the lochs were not so uniformly extended, and the area more of a bog dotted with pools of water. Among the tombs, the most important is the Maeshowe mound, 8 m high and 35 m wide. Inside runs a corridor built with slabs of stone weighing up to 30 tons each. It terminates in a large central chamber surrounded by three smaller cells. The scant human remains found in one of the cells allow us to date the tomb to around 2700 BC. Interestingly, the monument was surrounded by a ditch and raised bank similar to that of the henges. When filled with water, the ditch would then have had a function of "symbolic insulation" between the world of the living and that of the dead (Fig. 5.10).

The people who supervised the management of this complex ceremonial centre lived in Barnhouse, a neolithic settlement consisting of various buildings, one of which—built at a second stage—probably served a ritual purposes. Certainly connected with the activities of this area was also the village of Skara Brae, which

Fig. 5.10 Mainland. The Maeshowe mound from the south-west

lies a few kilometres to the north-west and probably was the main dwelling area of the priest-astronomers who were in charge of the sacred structures of Mainland. What is undeniable is that the place conveys a sense of austere, almost monastic order, on account of its meticulous "urban" planning.

This almost intact, concentrated series of monuments gives us some under-standing as to how the cosmic landscape of a Neolithic ceremonial centre was created. First of all, there is the issue of inter-visibility. The monuments indeed "speak to each other" through visibility lines which ideally criss-cross the whole area. In particular, at Stenness, a combination of slabs seems to form a window which, when looked at from the centre of the stone circle, pinpoints Maeshowe as a sort of magic portal connecting the two. Second, of course, there is astronomy. The corridor at Maeshowe is in fact correlated with the setting of the midwinter sun in a curious way. Looking from inside towards the western horizon, about twenty days before/after the winter solstice, the setting sun disappears behind the crest of a distant hill (Ward Hill, on the island of Hoy), then, however, reappears ("rises again") for some minutes at the base of the same hill before setting definitively. This beautiful phenomenon occurs because the sun's trajectory carries our star behind the hill's shoulder-like protuberance, interrupting the line of sight from Maeshowe; but if the day is sufficiently close to the solstice, the trajectory will be stretched sufficiently so as to make the sun visible again for a while. In a sense, then, the place where Maeshowe was built—probably on a pre-existing henge—was chosen because precisely there the sun would set and then rise again, giving both a relatively accurate determination of the solstice *and* a remarkable hierophany (Figs. 5.11 and 5.12).

Fig. 5.11 Mainland, December 1. The sun viewed from maeshowe sets behind Ward Hill at 3:02 pm (*Courtesy* V. Reijs)

Fig. 5.12 Mainland, December 1. The sun viewed from Maeshowe rises again for a few minutes at the base of Ward Hill at 3:09 pm (*Courtesy* V. Reijs)

When viewed from behind, with the hills at the horizon, also the shape of the Maeshowe mound in itself seems to recall that of the distant mountains to which it is astronomically connected. Perhaps this is a seeming effect connected with the present state of the mound, but in any case there is no doubt that a key element of the cosmic landscape of Mainland is the direct replication of nature. Indeed, as we have seen, the ditches of the henges were excavated in the bedrock and therefore functioned as natural pools; it is very likely that they remained filled with water for most of the year. The result—large mounds and circular banks surrounded by rings of water—is actually an imitation of the surrounding landscape that is, the promontory of Stenness itself (Richards 1992, 1996).

The second example worth discussing here is the urban planning of one of the largest cities in Precolumbian history, Teotihuacan, today in the outskirts of Mexico City. The city underwent a boom in the first centuries AD, to the extent of becoming one of the largest cities on the planet at that time (Millon and Dewritt 1975). The influence of Teotihuacan art and culture on the whole of Mesoamerica was enormous; it endured for some five centuries, until the city vanished from history for no clear reason. It then became a mythical place to the extent that the Aztecs—whose empire flourished many centuries later—gave it the name by which we call it today, which means "the place where the gods were born" (Fig. 5.12).

The urban plan was conceived from scratch by an evidently meticulous architect and it is centred on a main road, 2.5 km long and 40–95 m wide, the so-called Avenue of the Dead. Two large pyramids dominate the view, one at the northern end of the avenue, the other on its east side: the so-called Pyramid of the Moon and Pyramid of the Sun. The latter is one of the largest in the world. Its base originally had a 215 m long side, later extended to 225 m, and it was 63 m high (today it is made up of five platforms, but one of these is due to a shameful restoration). The Pyramid of the Moon, smaller but equally elegant, faces onto a large square and perfectly concludes the path along the Avenue. This path contains the key to understanding the cosmic landscape ideas which governed the planning of this sacred place (Broda 1993, 2000). The Avenue of the Dead is in fact aligned to a mountain, the Cerro Gordo, to the north. One only has to walk along the Avenue to realise that the silhouette of the Pyramid of the Moon in the distance seems to blend with the profile of the mountain: it is as though someone had made a scale copy of the mountain itself. This intuition is confirmed when it dawns on one that, from the northern end of the road, the gaze towards the huge mole of the Pyramid of the Sun is occupied in the distance by the profile of another mountain, Cerro Patlachique, located in the background, and that the slopes and the platforms of the pyramid were inspired by the profile of this mountain (Figs. 5.13, 5.14 and 5.15).

We can see, then, that the planners of Teotihuacan wanted to *replicate the mountains* in their cosmic landscape. Teotihuacan is therefore one of the most extraordinary examples of a sacred landscape, a place where a monumental replica of nature conferred power and prestige. Of course, we would expect to find the presence of astronomy there as well and, indeed, the archaeoastronomy of Teotihuacan has been widely studied. Here I shall mention only one principal fact. The city is among the first examples of (and perhaps the one which was of

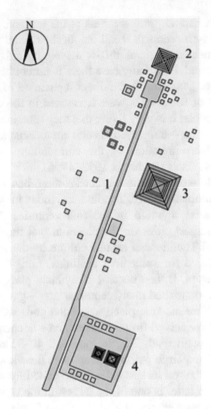

Fig. 5.13 Schematic plan of Teotihuacan: 1 Main ceremonial road or "Street of the Dead". 2 Pyramid of the Moon. 3 Pyramid of the Sun. 4 Ciutadela temple

inspiration for) an entire family of Precolumbian towns ("the 17° family", as Aveni and Hartung called it) whose main axis is skewed in relation to true north by an amount close to 17° (at Teotihuacan it is 15° and a half). The orthogonal azimuth, and thus essentially that of the Pyramid of the Sun which is parallel to the street, points to the setting of the sun on April 30 and August 13. These days have no special significance for the cycle of the sun at Teotihuacan (in particular, the two zenith passages occur much closer to the solstice at that latitude) but are separated by 260 days, which is a number that recurs in Mesoamerican calendars (as in the Maya Tzolkin, see Sect. 9.1) and in typical Mesoamerican orientations as well. A practical function can also be detected since the same alignment when read in the opposite direction yields two dates (February 11 and October 29) which, together with the first two, can be used as a rough timing guide for important agricultural activities (Iwaniszewski 1991; Sprajc 2000, 2001).

A third striking example worth mentioning here is that of a "no-alternative" sacred landscape: a place which was *necessarily* sacred for the people who settled there, given that it is the most out-of-the-way place on earth: Easter Island. Lost in the immensity of Pacific Ocean, Easter Island is a tiny triangle of volcanic land

Fig. 5.14 Teotihuacan. View on the main avenue, with the so-called pyramid of the Sun to the left

Fig. 5.15 Teotihuacan. The so-called pyramid of the Moon

160 km^2 large, 2000 km from the nearest island to the west, and 3500 km from the Chilean coast, to the east. The island was settled by Polynesians in the first millennium AD (Van Tilburg 1994). The islanders probably realised that they were stuck there for good and turned the island into their own sacred landscape by means of a cult whose *raison d'être* was based on the quarrying, carving, transporting and erecting of huge stone statues called Moai. Their civilisation collapsed about one hundred years before the arrival of the Europeans in the eighteenth century, and therefore we know little about the Moai cult. In any case, the statues represent

Fig. 5.16 Easter Island. Moai on the slopes of Rano Raraku

individuals with oblong traits, embellished with bas-reliefs representing ears with
elongated lobes, clothes and other objects. The monoliths are made of rock from
one of the extinct volcanoes of the island, the Rano Raraku, whose sides were used
as an open air quarry. As many as 900 monoliths were mined here, and at least
some 300 were successfully hauled up to their final destinations, huge stone platforms called Ahu.

In spite of the difficulties in interpreting this enigma, the vast majority of scholars
believe that the Moai pertained to a cult which involved the worship of ancestors,
represented by the statues. The idea seems convincing, as most of the platforms are
near the shoreline; however, the statues do not look out to the sea but rather inter-
nally, towards the settlements and the interior of the island. Thus, they form a sort of
ideal border, necessary for delineating and therefore legitimising the (compulsory)
life on the island. The whole of Easter island became in this way a sacred landscape,
and, as such, familiar mechanisms can be seen in action. The main one—in my
opinion—is the symbolic replica of natural features (Magli 2009). Indeed, when seen
in profile, the Moai resemble quite clearly the profile of the volcano Ranu Raraku,
the quarry where all the statues come from. Ranu Raraku itself is a sort of Moai
nursery: statues of different dimensions and at different stages of production emerge
from the rock, some abandoned, others ready to be transported or simply standing
there and left in place on the slopes of the volcano, gazing wistfully outward. The
place is so crowded by Moai, each so similar to each other, and all so similar to the
profile of the mountain, that it seems that the mountain is *cloning* itself, replicating
itself compulsively on a smaller scale and, with minimal variations and with various
levels of definition, in a obsessive sequence that reminds M.C. Escher's famous
lithographs (Magli 2009) (Figs. 5.16, 5.17 and 5.18).

As is to be expected, of course, also astronomy plays a role. In fact many Moai's
platforms are parallel to the shoreline, but many others show a clear tendency to be

Fig. 5.17 Easter Island. The profile of Rano Raraku

Fig. 5.18 Easter Island. The profile of a Moai (picture rotated 90° to facilitate comparison with the previous picture)

aligned with astronomical phenomena, in particular to the rising/setting positions of stars—the Pleiades and Orion's Belt—which marked seasonally significant moments of the year (Belmonte and Edwards 2004).

To conclude, then, in the above mentioned examples, as well as in many others we shall meet in Part III, the cosmic landscape is built up from a series of different elements—partly geographical, partly architectural, partly astronomical—which together contribute to creating scenarios of impressive complexity. There is usually no universal key able to disclose the meanings of such a complexity. Archaeoastronomy however—taken in the broadest sense and within a multi-disciplinary context—plays a fundamental role in giving us insights. The next chapter will develop this global approach as the basic scientific foundation of the discipline.

References

Belmonte, J. A., & Edwards, E. R. (2004). Megalithic astronomy of Easter Island: a reassessment. *Journal for the History of Astronomy, 35*(4), 421–433.

Broda, J. (1993). Astronomical knowledge, calendrics, and sacred geography in ancient Mesoamerica. In C. L. N. Ruggles & W. J. Saunders (Eds.), *Astronomies and cultures* (pp. 253–295). Niwot: University Press of Colorado.

Broda, J. (2000). Calendrics and ritual landscape at Teotihuacan: Themes of continuity in Mesoamerican cosmovision. In D. Carrasco, L. Jones, & S. Sessions (Eds.), *Mesoamericàs classic heritage: From Teotihuacan to the Aztecs* (pp. 397–432). Boulder: University Press of Colorado.

De Boer, J. (2014). The Oracle at Delphi: The Pythia and the pneuma, intoxicating gas finds, and hypotheses. In P. Wexler (Ed.), *History of toxicology and environmental health. Toxicology in antiquity I.* Amsterdam: Elsevier.

Devereux, P. (2010). *Sacred geography: Deciphering hidden codes in the landscape.* London: Gaia.

Dronfield, J. C. (1995). Subjective vision and the source of Irish megalithic art. *Antiquity, 69,* 539–549.

Durkheim, Emile. (1965). *The elementary forms of the religious life.* New York: Free Press.

Eliade, M. (1959). *The sacred and the profane: The nature of religion.* London: Harcourt.

Eliade, M. (1964). *Shamanism: Archaic techniques of ecstasy.* Princeton: Princeton University Press.

Eliade, M. (1971). *The myth of the eternal return: Or, cosmos and history.* London: Bollingen.

Iwaniszewski, S. (1991). La Arqueologia y la Astronomia en Teotihuacan. In J. Broda (Ed.), *Arqueoastronomía y etnoastronomia en Mesoamerica.* Mexico: Universidad Autonoma de Mexico.

Krupp, E. C. (1997). *Skywatchers, shamans, and kings.* New York: Wiley.

Magli, G. (2009). *Mysteries and discoveries of archaeoastronomy.* New York: Springer.

Millon, R., & Drewitt, B. (1975). *Urbanization at Teotihuacan.* Austin: University of Texas Press.

Richards, C. (1992). Barnhouse and Maeshowe. *Current Archaeology, 31,* 444–448.

Richards, C. (1996). Monuments as landscape: Creating the centre of the world in Neolithic Orkney. *World Archaeology, 28*(2), 190–208.

Schele, L., & Freidel, D. (1990). *A forest of kings: The untold story of ancient Maya.* New York: William Morrow.

Sprajc, I. (2000). Astronomical alignments at Teotihuacan. *Latin American Antiquity, 11,* 403–415.

Sprajc, I. (2001). *Orientaciones astronómicas en la arquitectura prehispánica del centro de México.* México: Instituto Nacional de Antropología e Historia (Colección Científica 427).

Steele, J. M. (2007). Celestial measurement in Babylonian astronomy. *Annals of Science, 64,* 293–325.

Turner, V. (1969). *The ritual process: Structure and anti-structure.* Chicago: Aldine Publishing Company.

Van Tilburg, J. (1994). *Easter Island: Archeology, ecology, and culture.* Washington: Smithsonian.

Chapter 6
The Scientific Foundations of Archaeoastronomy

6.1 Archaeoastronomy as a Cognitive Science

In this chapter we shall investigate the place of archaeoastronomy within archaeological research, aiming to assess its methodological affinity with other scientific and humanities disciplines. Our objective is also to understand what the most desirable compromise between a rigorous scientific approach to data and the peculiar, human-dependent nature of archaeoastronomical records might be.

As a starting point, notice that it is difficult to think of a scientific discipline which does *not* have a role to play in archaeological research today. A complete study of a site nowadays might well involve botanic as well as forensic pathology, require geo-radar investigation and petrographic analysis, and so on. All these disciplines have, of course, a flourishing life on their own, but—as is natural—a scientist usually feels value has been added to his work if it can furnish information in a fascinating field such as the study of human past. What about astronomy? Is archaeoastronomy an ancillary field? The answer is, at least in my view, that archaeoastronomy *can* be something much more than an ancillary, or secondary, discipline. If approached correctly it can actually be a powerful key to deciphering the thought of the past, through a general approach that is known as cognitive archaeology.

Cognitive archaeology is an attempt to construct a theoretical framework for archaeological data (Renfrew and Zubrow 1994). The output of such a framework aims at going beyond materiality and supplying a model of the way objects and environments were perceived in the minds of the people involved with them, and of the symbolic value they may have attached to them. Here, objects must be understood in the broadest sense: an "object" could be a small ceremonial axe or a pyramid weighing eight millions tons. Since human cultures collaborate in constructing a shared worldview of the cosmos, the cognitive approach should, at least in principle, be applicable not only to single experiences—testified to by objects or remains—but also to complex architectures and even to entire landscapes

© Springer Nature Switzerland AG 2020
G. Magli, *Archaeoastronomy*, Undergraduate Lecture Notes in Physics,
https://doi.org/10.1007/978-3-030-45147-9_6

developed over the course of centuries, with admittedly ambitious objectives in mind, such as understanding economics, politics, ideology, religion, and the management of power (Hodder 1999). An example may help in clarifying these ideas. As we shall see in details in Sect. 7.1, stone boulders of a quality called bluestone were apparently brought to the megalithic site of Stonehenge from as far away as the Preseli mountains (Wales), a distance of more than 200 km. Well, an exact science, petrography, can confirm for us that the Stonehenge bluestones really did come from Preseli, in spite of the fact that good quality stone was readily available much closer at hand. Another exact science, physics, can provide C14 dating of organic remains found in connection with the megaliths and so can tell us approximately when their transportation took place. Mine engineering can tell us how the Preseli stones were quarried, and experimental trials based on physical feasibility can give us an idea of the likely methods of transport. An archaeological survey can come up with numerous details regarding the extent to which the Preseli mountains were frequented by human society in the Neolithic period. A processional archaeology analysis can address the economic and social conditions that made such transportation possible.

But only a cognitive approach has any hope of telling us *why* they embarked on such an apparently crazy endeavour.

In a paper published in the 1970s, Alice and Thom Kehoe described the cognitive approach as follows: "the archaeologist must approach his data with the expectation of describing concrete objects that in reality had their primary cultural existence as percepts in topological relation to one another within the cognitive schemata of human beings." (Kehoe and Kehoe 1973). In other words, archaeological remains must be contextualised into a framework that is wide enough to include the time and location of their conception and construction, taking into account historical, climatic and economic conditions, the physical environment and the landscape, the symbolic environment and the cultural landscape, as well as the knowledge, mentality and religious beliefs of the builders. Ambitious as it may seem, a cognitive approach to archaeological reality should be all-encompassing, *even if the aim is to understand perhaps only one specific aspect.*

The specific focus of interest of archaeoastronomy is the perception and the contextualisation of the celestial cycles in material objects, and it is therefore clear that archaeoastronomy is a cognitive science. The idea of adopting a sweeping comprehensive approach towards data is—at least in my view—both the blessing and the curse of archaeoastronomy as a modern cognitive discipline. We must approach problems from a global angle even when the objects to be considered as "percepts in topological relation to one another" are as huge as the pyramids of the fourth Egyptian dynasty at Giza (Sect. 8.2) or the Pantheon (Sect. 10.4) and our aim is "only" to understand if they were related in some way to the stars.

It goes without saying, of course, that in cognitive archaeology there is the risk of introducing ideas, parallels and interpretations which may come from the experimenter's own experience and cultural background, and may have little or even nothing to do with the original purposes of a piece of art or even, in the worst scenario, with an entire archaeological site. An example may be of assistance.

In the Odyssey, Ulysses visits the Kingdom of the Dead. To find the entrance, the sorceress Circe advises him to search for a place where the two rivers Pyriphlegethon and Cocytus flow into in the Ackeron river. This place is located in Epirus, the north-western region of Greece, not far from the mouth of the Ackeron. Here, a hill was identified—at least from the fourth century BC if not before—with the site of the entrance to the "underworld", as described by Homer. On the hill, historical sources tell us that a sanctuary to Persephone and Hades, gods of the underworld, was founded. The sanctuary was famous for its oracle of the dead, or *Nekromanteion*, which allowed the pilgrims to enter into contact with the souls of the dead. On this very same hill, in 1958, the Greek archaeologist Sotirios Dakaris discovered an imposing building which had been re-utilised as foundations for a monastery (Dakaris 1993) (Fig. 6.1).

The building consists of a massive square enclosure in refined polygonal stonework. The arrangement of the rooms inside is complex and labyrinthine; the complex also boasts a large underground vaulted hall. Naturally, Dakaris identified this monument with the Nekromanteion, also on account of descriptions given by ancient authors, who speak of a visit to the oracle of the dead as a truly unforgettable experience. In fact, those who intended to consult the oracle were first isolated for a few days in dark rooms (identified by Dakaris in the northern sector of

Fig. 6.1 Nekromanteion. The underground vaulted room interpreted as the oracle's seat

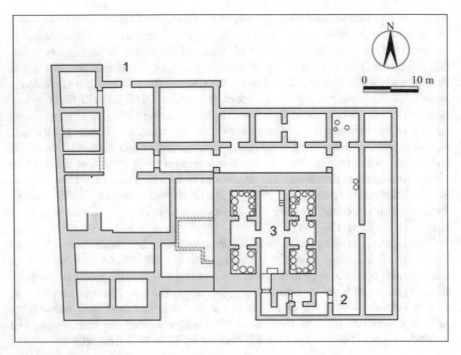

Fig. 6.2 Schematic plan of the Nekromanteion *1* Entrance to the polygonal precinct *2* Labyrinth *3* Main room with staircase to the underground vaulted room

the building). Here they followed a special diet and participated in rituals during which priests recited incantations and invocations, with the aim of preparing for the encounter with the dead. Subsequently, the pilgrim would probably pass through other rooms in the semi-darkness, finally entering the subterranean hall of the oracle. Among the archaeological finds supporting this view are toothed wheels of bronze, which Dakaris interpreted as gears used for a lifting mechanism by means of pulleys. Such machines were perhaps installed in the hall and used for dramatic effects, making the meeting with the departed all the more spectacular and gratifying (Fig. 6.2).

Well, incredible as it may seem, after Dakaris' discovery a *completely*—I repeat, completely—different interpretation of the very same place was formulated (see Wiseman (1998) and references therein). Indeed, the wheels actually appear to be mechanisms of catapults, and consequently some authors believe that Dakaris had dramatically misinterpreted the building, which according to them was rather a *fortified farm*. The subterranean hall is the farmer's cistern or a storage room, and the fortified place was destroyed in a skirmish occurring during the Roman conquest of 167 BC.

Attempting to establish which of the two interpretations is more credible is somewhat problematic and would lead us away from our main thread of thought.

Fig. 6.3 Helleniko. The Pyramid

For the interested reader I only mention that the orientation of the Nekromanteion is, with a good approximation, cardinal, an orientation which is hardly justifiable for morphological—or even less functional—reasons, and that the use of polygonal stonework in many cases signals a symbolic destination of the buildings; this occurs in Greece, for instance, in the case of the Pyramid of Helleniko (Fig. 6.3) for which many problems of interpretation and dating occur (see Fagan 2006; Theocaris et al. 1996), and for the many monuments in polygonal stonework of Central Italy (Magli 2007). In any case, and whatever the truth about the building discovered by Dakaris may be—one of the most important oracles of the dead of the Classical World, or a fortified Hellenistic farm—this example shows the broad risks attendant on a cognitive interpretation of archaeological data.

In the case of archaeoastronomy, this concern arose already in the sixties of last century, even before the development of cognitive archaeology itself. Indeed in 1963 Gerald Hawkins, a professor of astronomy at Boston University, ventured to apply a (for those times trailblazing) computerised analysis of the possible astronomical use of stones' alignments and pits at Stonehenge. The results (first published in *Nature* and later in the book *Stonehenge Decoded*) showed Stonehenge to be ultimately a "Neolithic computer", capable not only of observing the cycles of the Sun and the Moon on the horizon, but also of predicting eclipses. The resonance of the work was staggering, but the reaction of the world of Archaeology was frosty; the builders of Stonehenge were slammed by the authority Richard Atkinson as "howling barbarians", thus denying them any chance to develop their "megalithic science". Another comment that gained notoriety was made by Hawkes (1967) who wrote that "every age gets the Stonehenge it deserves—and desires".

Today, we do know much about Stonehenge (Sect. 7.1). We do know that Stonehenge was *not* a computational machine for astronomical cycles, as Hawkings had dared to suggest, and that its alignments are lacking in precision and could be used only symbolically, not for accurate astronomical observations. And we *do know* that astronomy was present there, and was there almost from the very beginning. So again, archaeoastronomy is not the universal key, the passe-partout that can divulge all the secrets of such a complex monument. But it is, in many cases, an invaluable tool. Therefore, archaeoastronomy does not deal with, and does not lead to, discoveries that the archaeoastronomer deserves—or desires.

On the contrary, in many cases archaeoastronomy—taking into account in a judicious way the risks which are intrinsic to the cognitive approach—can lead us to uncover much of what a monument really has to say about the mindsets and the traditions of its builders.

6.2 Archaeoastronomy as an Exact Science

Archaeoastronomical research is based on a set of experimental data, as would be the case in any other science. Data processing is then performed by using a set of scientific instruments. Within what limits can we speak of archaeoastronomy as being an exact science, and what are the most common errors deriving from these limits?

The scientific approach to any set of experimental data hinges on a fundamental characteristic of any physical experiment: repeatability. A theory is considered as proved if all repeated experiments, under the same conditions and estimated errors, give the same results (it is of course assumed that any further similar experiment will give the same result in future). An archaeological experiment—a dig—is, on the other hand, not repeatable by definition. Modern archaeologists know this very well and try to be as complete as possible—they will usually leave unexcavated areas for future "experiments" as well. In archaeoastronomy, the problem associated with non-repeatability is dramatically reflected in the size of the available data samples. It is rare indeed to have dozens of measurable monuments of the same type, not to say hundreds of them. The implication of this is that only a very cautious use of statistics can be made (Sect. 3.4). Moreover, as in any other scientific discipline, experimental results in archaeoastronomy can be influenced by the experimenter (we are speaking here only of involuntary, and non fraudulent, influences, of course). In archaeoastronomy, this typically results in a "selection effect". To avoid it, the maximum possible amount of information about the experimental conditions must always be supplied.

Even if the method of acquiring data used is technically correct, there are still errors to which the (let us say) negligent archaeoastronomer may fall prey. These errors typically arise in the phase of processing the data and comparing them with astronomical data. The result of this kind of errors is that the web is full of sites where a supposed archaeoastronomical deduction is used to validate the most fanciful, not to say ridiculous, theory. Every day new theories pop up, and I could mention at least five different astronomical explanations for the arrangement on the ground of the pyramids of the Giza plateau (a topographical enigma which, as we shall see in Sect. 8.3, is relatively easy to solve if the cognitive approach is applied). The most common drawback of such theories has to do with chronology and thus, ultimately, with deficiencies in historical and archaeological background. It could be defined as a posteriori selection of the data, in parallel with the selection error in field, which is a priori treatment of the data. As an explanation I shall recall a notorious example, which—in spite of having repeatedly been shown to be nothing but a childish misunderstanding—is still quite popular on the web and also crops up on televisions.

In the northern regions of Bolivia near Lake Titicaca there flourished the culture that was named after the monumental centre of Tiahuanaco. As a matter of fact, Tiahuanaco was the largest and most important city in the history of the Andes before Cusco, the capital of the Incas, and can be relatively well dated to between the third and the sixth centuries AD. The monumental centre of the city is dominated by a precinct called Kalasasaya, a rectangular platform 130 × 118 m wide, constructed with massive blocks. One of the first published works describing this place we owe to Arthur Posnansky (1873–1946). Posnansky was born in Vienna, Austria. After graduating as a naval military engineer, he emigrated to South America, first to Brazil, and then Bolivia. In his travels he studied and published several books on Bolivian archaeology and geography. In particular, he was much impressed by Tiahuanaco and his book on this site remains as an important record, which contributed to the dissemination of knowledge on this civilisation and to the preservation of its monuments (Fig. 6.4).

Fig. 6.4 Tiahuanaco. The Kalasasaya platform

In his highly accurate maps, he deduced that the Kalasasaya is well-orientated to the cardinal points, the average error being less than one degree. This is, of course, important information that he gleaned from this monument. However, Posnansky went further. He noticed that from the midpoint of the west wall of the platform, looking towards the north-east and at the south-east corners, the azimuth is of 24° respectively north/south of east. This corresponds remarkably well with the azimuths of the rising sun at the two solstices at the latitude of Tiahuanaco, the error being around 40′ in total. Therefore it might be supposed, as Posnansky did, that the dimensions of the platform were fixed according to this astronomical criterion. Again, this is an interesting observation, and any Archaeoastronomer today would be more than satisfied with these results. For some reason, Posnansky was not. He became convinced that the errors were too large, bearing in mind the capacities of the Tiahuanaco architects. So, he ventured to propose that the skew of the platform in relation to the cardinal points was a later displacement due to some (admittedly huge) earthquake-like force, and that the error committed by the builders in their solstice orientations was virtually *zero*. Given such premises, it was clear that it was the sun that was in the wrong place at the solstices, not the alignments. Accordingly, he went back to the sky to ascertain if there existed a date which fitted in with his presumed zero error or, in other words, turned the sky back in time until the sun was kind enough to rise in perfect alignment at the two solstices. As he knew and we know, the azimuths of the sun do not depend on precession, but the very slow change of the obliquity of the ecliptic provokes a small change in the maximal declination and therefore in the azimuths. This finally gave Posnansky the desired correction. The corresponding date was, according to him, proof that the place was constructed "before the deluge", approximately 15,000 BC (Posnansky 1945).

Posnansky's analysis is technically correct but, of course, his results are ridiculous, in that they are based on a macroscopically wrong assumption, namely that the builders committed no errors. As a consequence, his conclusions plainly contradict the archaeological evidence. The error made by Posnansky is quite glaring, and continuing to repeat it today is nothing short of outrageous (in spite of this, his ideas on Tiahuanaco are still being trundled out as proof of a supposed Atlantis civilisation). More generally speaking, there is practically no doubt that the measurement of solar alignments *cannot* be used as a dating method, and attempts to do this have to be regarded with great suspicion. More delicate is the case of star alignments. Here precession comes into play and adjusting, by precession, a measured alignment to the azimuth of a suitable star can require only some cen-turies of allowance. Thus, yet again, it is a risky procedure. However, this does not necessarily mean that stellar astronomical dating *cannot* be used: as we shall see in Sect. 8.1, for instance, Archaeoastronomical investigation of the Great Pyramid furnishes a brilliant, objective proof of its archaeological dating according to Egyptology. The pivotal point is thus again seen to be the interplay with the archaeological context and, if necessary, with other possible sources of information.

Yet another important issue regards astronomical dating and selection. There were more than one thousands stars visible to the naked eye on a clear night anywhere in antiquity (or in a pollution-free, light-pollution-free location today).

If we have any azimuth at a specific place, it is relatively easy to discover a star rising or setting to that azimuth, especially if a wide error is allowed and/or the century can be adjusted using precession. Clearly then, claiming that a corresponding alignment was deliberate requires a careful analysis, which must take into account, first of all, that only on extremely special occasions can a low magnitude star be the objective of an alignment (examples exist: one is certainly the Pleiades in the Precolumbian world, which attracted enormous attention due to their special nature, although their global magnitude is over three; another is the faint constellation of the Dolphin in the Greek world). Thus possible astronomical targets usually have to be restricted. Then a solid cultural basis must be established for the claimed alignments. For instance, the interest in the Pleiades on the part of the Incas is documented by several Spanish sources as well as ethnographically, while Dolphin was one of the constellations associated with the myth connecting the God Apollo with his own sanctuary at Delphi. Only if these conditions are satisfied, the possibility of a stellar alignment should not be rejected.

I conclude this section by noting that, sometimes, the overriding need to prove the intention behind alignments has been seriously misguided also on the part of archaeologists. One may happen, indeed, to hear or even to find in written literature that this is a *structural drawback* of archaeoastronomy itself. This is of course quite absurd: any scientific discipline is fraught with a number of characterising difficulties, and any serious scientist is aware of the shortcomings in his own discipline and does his best to overcome them, hopefully without non-experts simultaneously attempting to throw the whole subject into disrepute. Again, an example might be of help.

The Nasca drainage plain is a vast semi-deserted area in the southern part of Peru. Although it rains very little in Nasca, the ground is relatively fertile and there are seasonal rivers, subterranean folds and springs. Nasca was the cradle of a civilisation, dated between the second and sixth centuries BC, and had its main ceremonial centre in Cahuachi. In 1920 the pilot of an aeroplane noticed that the Nasca plateau is crossed by an incredible number of artificial lines. In doing so, he unveiled one of the most intriguing mysteries of Pre-Columbian archaeology. On the deserted surface, in fact, the Nasca people "drew", by digging superficial tracks on the ground and removing the pebbles from the inside of the tracks, thousands of kilometres of lines. Most of these etchings on the desert surface run straight; others create huge trapezoidal figures and finally a few form gigantic drawings of living beings, dozens or even hundreds of metres long, called zoomorphic geoglyphs (Fig. 6.5). The Nasca plain is a sort of gargantuan work of art, created over many centuries. But why? In the 1950s the explorer Paul Kosok and the mathematician Maria Reiche put forward the idea that Nasca was a sort of giant astronomical treatise; they actually found a few lines and figures which showed likely astronomical orientations but, as it turns out, they never found the universal key to back up their treatise theory (Reiche 1980). The astronomical hypothesis of Kosok and Reiche was re-considered in 1968 by Gerald Hawkins, who performed an archaeo-astronomical study of the site on behalf of the National Geographic Society. Hawkins' study was based on a small number of lines, for a total of 186

Fig. 6.5 Nasca. One of the zoomorphic geoglyphs. This one resembles a hummingbird

directions. He proceeded with a computerised analysis, using a similar method to the one he had pioneered a few years before at Stonehenge, and concluded that "the ancient lines in the desert near Nasca show no preference for the directions of the sun, moon, planets, or brighter stars…the pattern of lines as a whole cannot be explained as astronomical, nor they are calendric" (Hawkins 1969).

Today, we do have a somewhat clearer understanding of the Nasca lines, also archaeologically speaking, and we do know that Hawkings was correct: astronomy definitively is not a universal key to the Nasca lines. There are a few astronomical elements of interest, and the zoomorphic geoglyphs remain tempting as possible depictions of single constellations (their uniqueness is really striking, there is not even one figure repeated), but the multitude of straight lines represents a very complex cultural phenomenon mainly relating to processional routes and cere-monies, as well as, probably, to underwater flows (Aveni 1991; Johnson et al. 2002). Thus, a proper scientific approach adopted within a proper scientific discipline, archaeoastronomy, culminated in a correct result, which in this case happened to be negative. However, this is not always the way in which Hawking's work has been understood in archaeological literature. For instance, one can read that "an alignment between a celestial object and a ground marking is statistically insignificant because countless stars are visible in the clear night sky at Nasca" (Silverman and Proulx 2002), a statement that, I believe, the reader will be able to comment by himself.

6.3   Archaeoastronomy and Unwritten Sources

In the previous section we touched upon a core issue regarding the "exactness" of archaeoastronomy. To recapitulate, consider, for example, the macroscopic error of the Tiahuanaco dating by Posnansky. From the technical point of view, his analysis is sound: the reliefs of the site are precise, the north appearing on the relief is correct, the azimuths are carefully taken, the analysis of the ancient sky is correct. In spite of this, the conclusions drawn are absurd, as soon as one looks at the archaeological evidence. This means that archaeoastronomy can work if, and only if, scientific methods are combined with information and data from her "humanities" counterpart (Ruggles and Saunders 1993; Iwaniszewski 2005; Sims 2011). The principal aim of this synergy is to confirm that the builders' alignments at a site were indeed intentional.

This necessary interaction is by no means easy to achieve. First of all, any problem should be approached without a priori objectives. The main drawback of attempts that can be collectively defined as "pseudo-archaeology" (Fagan 2006) lies in fact in the very approach to the question: first one becomes convinced that something must be true, and then tries to prove precisely that thesis, in many cases selecting only the facts that fit the theory. Moreover, a serious difficulty with which archaeoastronomy is constantly challenged is the lack of written sources, that is, the absence of historical documentation. Naturally, it would be wonderful to discover a Neolithic monument astronomically oriented *together with* a document by the builders explaining the deliberateness and meaning of such a orientation but unfortunately they did not have the written word at all. It may instead well happen that a previously unknown ancient text is discovered about, say, the construction of the Pantheon in Rome. But up to that day, the only text available on one of the most important monuments of the Roman world is a couple of ambiguous statements made by the late Roman historian Cassius Dio.

This lack of written sources leads to a substantial lack of *absolute* evidence: an assertion in archaeoastronomy can usually be made only with a certain degree of confidence. This unfortunately risks denying validity to the discipline as a whole, in a prejudicial—and of course unacceptable—way. As a matter of fact, for instance, the entire development of the history of astronomy in the Mediterranean Basin, and especially in Egypt, has been plagued for decades by such prejudice, summed up in the works of an influential author, Otto Neugebauer. In fact, the idea that knowledge can only be handed down through the *explicitly written* word, along with the concomitant notion that the only knowledge that can be passed on orally is of a mythological, or broadly historical, nature (traditions, lives of the ancestors, etc.) even prompted him to say that "Egypt has no place in a work on the history of mathematical astronomy" (Neugebauer 1975). Together with Richard Parker, he published the entire corpus of what they considered to be the known astronomical Egyptian texts (Neugebauer and Parker 1964). Besides many other drawbacks which would take too long to discuss here, these books do not contain a single word about one of the most astonishing period of human history, the Old Kingdom of Egypt, during which refined astronomically oriented buildings were constructed and from which writings with a plethora of references to sky, the Pyramid Texts, have passed down to us (Sect. 8.1). Neugebauer and Parker thus failed completely to attach any value to astronomically anchored architecture, and failed to recognise any astronomical content in the Pyramid Texts, which are predominantly religious in nature.

On the contrary, myth has been advocated as the main source for ancestral astronomy in a fairly influential book by de Santillana and Von Dechend, *Hamlet's*

Mill (1969). These authors were convinced that a series of common patterns tending to recur in the mythology of many different cultures correspond to the description of the same celestial events, and in particular, to the shift of the constellation hosting the spring equinox (Sect. 1.6). So far, however, these ideas have not been independently confirmed, and, specifically, evidence of a discovery of precessional effects prior to Hellenism is mostly circumstantial (the most convincing case is in the context of ancient Indian astronomy, Kak 2000). Nevertheless, an analysis of the possible astronomical content of myths might well offer a fruitful field for future research.

Apart from ancient literature and historiography, there is yet another discipline which interacts significantly with archaeoastronomy. It is not infrequent for the cultural traditions of contemporary people to stretch far back in time to their remote ancestors, and in several cases this holds for astronomical lore too. As a consequence, ancient astronomy—and its connection with ancient architecture—may be investigated through anthropology, by studying the astronomical lore of modern peoples. One of the risks to this approach, however, is that it is sometimes difficult to be certain that existing traditions and knowledge really go back as far as it is assumed, but in some cases there can be no doubt whatsoever, as I shall illustrate with an amazing example (Fig. 6.6).

In historical texts written in Spanish in the sixteenth century, soon after the conquest of the Inca empire (Sect. 9.3), the presence of "dark animals" in the sky is mentioned as one of the characteristic of Inca astronomy. These texts had always baffled any attempt at interpretation until the end of the last century, when Urton (1982) published his ethnological study on the Misminay, a *present-day* people living in a village a few miles from Cusco. The Mismiminay see in the sky a series of animals (a llama, a baby llama, a fox, a toad, and others) located in the dark

Fig. 6.6 The southern part of the Milky Way as seen looking south in 1400 AD. The two branches, enclosing several dark areas, are visible. In these areas the Incas identified "dark" constellations, all representing Andean animals

zones of the Milky Way which correspond to our constellations Sagittarius, Scorpio, Southern Cross, Vela and Puppis. These people arrange these images into a band of dark constellations crossing the sky, in a sense similar to that of our zodiac, but dark and distributed on the Milky Way instead of the ecliptic. The animals corresponding to these images were both common and significant in Inca times, either for practical reasons (such as the llamas) or for reasons connected to myths and beliefs (such as the Fox and the Serpent) and there is little doubt that the tradition surrounding these dark constellations stems directly from the Incas.

In this example, ethnoastronomy has provided valuable input for the history of the Inca astronomy, enabling the interpretation of an obscure source to be made. When applied specifically to archaeoastronomy, however, ethnoastronomy should be used to identify, in the current interest in a celestial object, the cultural matrix of an ancient alignment—then things become more complicated. Again, an interesting example may make things clearer.

The Anasazi are the ancestors of the present-day American Indian tribes which live along the Rio Grande. Anasazi culture flourished around 1000 AD in the region which corresponds to the border area where Utah, Colorado, Arizona and New Mexico converge. The most important ceremonial site of the area is Chaco Canyon, in New Mexico. Here the Anasazi built a series of huge *Pueblos*: palaces containing several rooms and circular spaces (known as *Kivas*) dug into the ground and elaborately refined inside. The largest is the so-called Pueblo Bonito, which contains roughly 600 rooms and 36 Kivas. It is semicircular in shape, with a straight 150 m-long front. Inside, a wall runs perpendicular to the diameter and an enormous 17-m Kiva is set tangentially to the wall (Fig. 6.7).

Fig. 6.7 Chaco Canyon. Pueblo Bonito

The archaeoastronomy of the Anasazi has been recently subject to revision, which has somewhat weakened existing claims regarding the astronomical alignments of several buildings at Chaco and elsewhere. However, some facts are quite reasonable (Sofaer and Sinclair 1986; Malville and Munro 2011). First of all, the interest of these people in the cardinal points, as shown for instance by the very precise orientation of the longitudinal axis of Pueblo Bonito on the east-west direction and by the tracing of straight roads without any utilitarian function (probably used for pilgrimages), one of which runs close to the meridian. Second, the interest in the solar and lunar cycles. Lunar observation is, in particular, attested to at a very singular Anasazi site called Chimney Rock.

The Chimney Rock Pueblo is built on a desolate hill ridge that lacks water sources; it contains two Kivas and 35 ground floor rooms. It is difficult to explain the existence of this building for functional purposes. However, the landscape here is very special, as the horizon to the north-east is dominated by two spectacular rocks or "chimneys". From the east court of the Pueblo one has a perfect view of these Chimneys and one can even sit on a ancient bench here which seems to have been built precisely with this purpose in mind. As a consequence, there is the distinct possibility that the Pueblo was constructed, at least in part, in order to use the double chimneys as a foresight for the observation of astronomical phenomena. In particular, the moon close to maximal declination rises between the chimneys. The minimum required declination is about half a degree less than the declination at standstill. Consequently, the rising of the moon is visible for some three years every 18.6 years. Using dendrochronology (the technique which allows dating of wood in structures by analysing tree rings) it is possible to correlate lunar standstills with the chronology of construction of the Pueblo. Indeed, the initial construction of Chimney Rock was in 1076 AD, a year of major lunar standstill, and an enlargement occurred in 1093, immediately before the subsequent lunar standstill. As discussed in Sect. 1.7, there is virtually no doubt that observation of lunar maxima in antiquity was concentrated on the full moon, and therefore at Chimney Rock (where the major northern standstill was viewed) the most important astronomical event must have been the rising of the full moon between the chimneys near the time of winter solstice. The spectacle itself was (and is) quite impressive, and it must have been even more for the Anasazi, if—as has been suggested—the twin rock towers were perceived as a shrine devoted to a couple of Twin Gods associated with War, children of the sun and the moon. Thanks to such a distinct and extraordinary feature of the landscape, indeed, Chimney Rock might have developed as an independent (as opposed to Chaco) centre for ceremony and pilgrimage (Malville 2008) (Fig. 6.8).

To return to our problem, given the cultural continuity between the Anasazi and present day American Indians, it is an interesting exercise to compare their cosmology and lore of the sky, as investigated by anthropologists, with the only information we have on Anasazi astronomy: namely, that coming from archaeoastronomy. The cosmology of the Pueblo Indians, for instance, is replete

Fig. 6.8 Kimney Rock. The full moon rises between the Kimneys (*Courtesy* Kim Malville)

with roads and is clearly "oriented". Life is a road, and the spirits are custodians of the various roads that branch off from it. Humans first emerged from a hole in the floor of a Kiva, then moved southwards, and so, logically, the spirits of the dead travel northwards, returning to the womb of the Earth. The spirits then return each year to visit the living, using the roads that had led them north, and these roads are often described as being straight. Very long pilgrimages to sacred sites are documented, and in particular, among the Zuni, a pilgrimage is undertaken every four years at the summer solstice toward a lake believed to host the spirits of the dead. The lake is seen as the center, the navel of the world, a sacred point on which all cardinal directions converge. Another interesting point is that sun and moon play complementary roles in overseeing religious rituals and agricultural activities—the year begins, for example, with the first full moon following the winter solstice (Ellis 1975; Zeilik 1985, 1986). Finally, observational astronomy is also documented: among the Hopi, for example, quite complex solar "horizon calendars" are used, both for practical and ritual purposes (McCluskey 1977).

The astronomical lore of the American Indians is therefore of enormous interest, and gives us considerable insight into the astronomically-anchored architecture of their ancestors. However, it should be noted that no explicit trace of any knowledge of the 18.6 year cycle of the lunar standstills has been documented. The presence of vague or mutated traces of past knowledge in ethnological data is actually common and, in particular, calendrical traditions appear to filter down more easily than knowledge of astronomical cycles, to the extent that some contemporary Maya populations still use the "sacred" Tzolkin calendar of 260 days today without being aware of the origin of its number of days (this calendar is discussed in Sect. 9.1).

6.4   Archaeoastronomy and Reverse Engineering

With any human artefact, from a car engine to computer software, *reverse engineering* (or architecture) is that series of processes and methods which aims to extract knowledge and/or design information from it. Today of course it is mostly applied at industrial level, in order to figure out (and sometimes to rip off) the products of competitors. It is not as well developed a tool as it should be, however, in studying past architecture. One only has to think, for example, that besides some noteworthy exceptions—in particular, the works by Protzen (1985, 1993) on Inca stonework—the study of the construction methods of megalithic buildings has largely been left in the hands of enthusiastic volunteers, sometimes leading to truly embarrassing disasters (for instance, the "Millennium stone project" of 2000, meant to bring a bluestone at Stonehenge from the Preseli mountains using the alleged ancient methods; the travel of the stone ended prematurely in the river Avon). Yet, a few cases of important reverse architectural design have been inspired by urgent conservation projects—we might mention two jaw-dropping masterpieces of Italian Medieval/Renaissance art: the Leaning Tower of Pisa and the Brunelleschi Dome in Florence.

The archaeoastronomical analysis of a architectural project might well be considered as a type of reverse architecture. In my experience as an Archaeoastronomer, I have often noticed that this point is usually overlooked, though it may be of considerable importance. Indeed, in many cases the aim of introducing astronomical alignments into a project entailed tricky design solutions and substantial additional effort for the builders. Clearly, if on one hand understanding such activities gives us a more general picture of the methods and the ideas of the builders, on the other it can also strengthen astronomical hypotheses. This, in particular, can be of paramount importance in the so-called "testis unus-testis nullus" cases. This Latin phrase essentially means that one should never rely on one source alone: "one witness is no witness". In exact science, it is certainly so: an experiment which is not reproducible is not an experiment. But archaeoastronomy is a multidisciplinary science based on historical remains, which cannot be duplicated at will and which are, in some cases, simply unique.

A blatant example is that of the so-called "air shafts" of the Great Pyramid. As we shall learn in Sect. 8.1, the weightiest monument ever built on earth, the Pyramid of Khufu at Giza, is crossed by four narrow shafts, each having the section of a handkerchief, which run *diagonally* into the core of the monument for dozens of metres (Figs. 6.9 and 6.10).

Traditionally overlooked by Egyptologists, and/or interpreted as ventilation channels (for a tomb…), they are rather a fundamental gear in the complex symbolic machine aimed at assuring eternal life to King Khufu amid the circumpolar stars and the stars of Sirius and Orion. Subsequently we shall study in depth all the relevant archaeoastronomical aspects; what I am at pains to emphasise here is that these shafts are a formidable engineering feat, which had been never properly understood until the exploration carried out in the 1990s (Gantenbrink 1999). In fact, the courses of

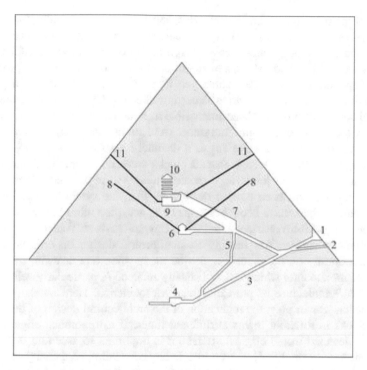

Fig. 6.9 Schematic section of the Great Pyramid. *1* Original entrance, *2* Looter's hole, *3* Descending corridor, *4*, Subterranean Chamber, *5* Well, *6* Queen's Chamber, *7* Great Gallery, *8* Lower shafts, *9* King's chamber, *10* "Relieving Chambers", *11* Upper shafts

Fig. 6.10 Giza. Detail of the King's Chamber in Khufu pyramid, completely built with huge granite blocks. On the right of the entrance (north) the mouth of the northern upper channel can be seen

blocks of the pyramid are horizontal, while the shafts run diagonally. This, first of all, means that a few metres of each shaft were added before the completion of each course, and after that the architect in charge of the shafts had to wait for the subsequent course to be laid, taking a matter of months, if not years, before proceeding with his operation, course after course, year after year, unable to commit any errors, since a displacement or a cracking occurring after the laying down of the superimposed course would have been impossible to remedy. Furthermore, to avoid the sliding of the blocks of the shaft, the stones were cut in a very singular manner. The shaft roof and both walls were cut as a channel in a single block; to guarantee stability against sliding, wedge-shaped blocks were used, and to avoid the displacement of continuous joints, appropriate solutions were devised. Clearly, if such technical aspects had been figured out before, then the symbolic function of the shafts would perhaps have been ascertained and accepted more easily, and without the (almost incredible) resistance that some Egyptologists still show.

Important considerations relating to architectural design on one hand and to archaeoastronomy on the other can also be made when one takes an overview of groups of monuments which are stylistically related. A particularly relevant case regards the architecture of pre-Columbian Mesoamerica. Here indeed a peculiar phenomenon was in play: the replication of the architectural design of buildings or entire blocks in different towns at different times. If astronomical alignments are plausibly detected in the original artefact, it is important to ascertain if they were also replicated, or modified, or even lost. In any case (all three actually occurred) it is equally clear that the role of archaeoastronomy is important for assessing the cultural continuity and/or change that characterises the replica (in the sequel to the present book we shall examine in detail a striking example in the relationship of the so-called Castillo of Chichen Itza with its replica at Mayapan, Sect. 9.2).

Yet another point—as far as the delicate connection between architecture, engineering skills and symbolism in ancient cultures is concerned—arises as a sort of chicken/egg dilemma. For instance, can we be so sure that religious shamanistic symbolism was at the root of architectural monumentality, or, at least in some cases, was it the reverse, that is, a doctrine was formulated and/or elaborated to account for the desire and the ability to construct monumental buildings? The suspicion arises, for instance, in Egypt, where the Texts of the Pyramids—supporting the idea of the voyage of the deceased Pharaoh into the sky—were put into writing, at least as far as we know, after those most giant machines devoted to such complex symbolic travelling, the pyramids of Giza, had been constructed. Also in the case of neolithic art carvings, which were probably inspired by altered states of consciousness, we have proof that many chambered cairns containing such carvings were second-hand constructions based on previous monuments and perhaps testifying to the wish to change, or elaborate the corresponding doctrine.

A final point where reverse engineering and archaeoastronomy intersect, and a point which is often ignored in archaeological studies, is that of the *location* of buildings. The selection of places for the construction of complex monuments was, in fact, in many cases, *not* dictated by practical, functional criteria but rather based on symbolic grounds. Again, the chief example is that of the pyramids of Giza,

Fig. 6.11 Rome. Castle S. Angelo, the former Hadrian's Mausoleum, and the river Tiber

which were placed exactly were they are—at enormous physical cost, for example, slicing 300 m off the rock plateau on the west side of the second pyramid—simply in order to follow the rules codified for that sacred landscape (Sect. 8.3). Another example is the mausoleum of Emperor Hadrian (later overbuilt as the Sant'Angelo Castle) in Rome, constructed around 120 AD (Fig. 6.11). The Mausoleum is built on the right hand side of the River Tiber, a place the ancient Romans did not particularly appreciate (the town developed on the left hand side), and this necessitated the construction of a brand new bridge. Furthermore, there was at that time still plenty of space in the Campus Martius, the area rearranged by Augustus as his own sacred space with the construction of his own mausoleum and of the first version of the Pantheon, which was later to be rebuilt by Hadrian himself (we shall discuss the Pantheon in detail in Sect. 10.4). So why did Hadrian make this choice? One explanation may be that in this way it was possible to place Hadrian's mausoleum along the azimuth of the setting sun at the summer solstice as viewed from the Pantheon, thus associating the apotheosised emperor with a powerful symbol of the divinity of Rome (Hannah and Magli 2011).

6.5 Towards Archaeology of the Cosmic Landscape

The discussion developed so far clearly confirms that archaeoastronomy is a far from subsidiary discipline. The potential of archaeoastronomy is noteworthy, given that the sky was an essential ingredient of almost any worldview, and because worldviews were reinterpreted and made concrete, using monumental architecture in cosmic landscapes of power. At any rate, in Part III we shall consider a series of case studies, taken from the most diverse civilisations and historical periods, which—I hope—will convince even the most sceptical of the readers.

Archaeoastronomy, then, can be seen as playing a seminal role in the cognitive analysis of countless sites, although this per se does not exhaust the potential of this discipline. Indeed, the basics of archaeoastronomy should be known also to modern architects, as astronomy can be seen in action also in contemporary architecture, adding a significant value to special places and monuments. One example is the memorial to the victims of the September 11, 2001 attacks, built in front of the Baltimore World Trade Center and inaugurated on September 11, 2011, the tenth anniversary of the attacks. The memorial, which is made of three original steel beams from the destroyed WTC north tower, is constructed on top of marble blocks bearing the names and birth-dates of the 68 Marylanders who died in the attacks. The position of the plinth has been designed in such a way that, on September 11, the shadow of the huge skyscraper moves across it like a sundial, progressively highlighting the inscriptions that commemorate those tragic events.

A striking and intrinsic feature of archaeoastronomy is the possibility of rediscovering spectacular concomitances relating to the sun, the moon and Venus which were fixed in ancient times and successively forgotten. This may culturally enhance the site in question and is potentially very important for the dissemination of archaeoastronomy itself among the general public. Moreover, and more generally speaking, it may play a relevant role in the conservation and promotion of cultural heritage. Of course, archaeoastronomical events—such as the midsummer sunrise at Stonehenge or the equinoctial hierophany at Chichen Itza'—must be correctly described and documented, but also carefully safeguarded to avoid any possible danger to the sites as well as to allow the public to enjoy them. In this respect, a potentially important initiative connecting archaeoastronomy with the field of conservation and positive fruition is that linking Archaeoastronomical Heritage with the Unesco World Heritage Convention (Cotte and Ruggles 2010).

More problematic is the possible role of archaeoastronomy itself as an unwritten source. In other words, to what extent can astronomical alignments be used to extrapolate information about the builders? As we have seen, archaeoastronomical dating through sun positions has to be ruled out as infeasible, and stellar dating can only be performed with the utmost care, essentially only to confirm the already established dating of a monument obtained by other means. However, the relationship between archaeoastronomy and reverse engineering can furnish us with quite valuable information about the choices behind projects and the ideas underpinning them.

To conclude, our understanding of ancient built landscape on a cognitive level has made great advances in recent years, and archaeoastronomy has given, and can give in the future, an important contribution to this global archaeological knowledge. This occurs both in the cases where the sky is a key element of the cosmic landscape, and also where it is not, provided that cultural (topographical, dynastic and so on) motivations are seen to be in play. Archaeoastronomy is now a universally accepted nomenclature and it would be absurd to try to change it now. However, this discipline has evolved substantially since it first saw light, and a more accurate name might be *Archaeology of the cosmic landscape.*

References

Aveni, A. F. (1991). *The lines of Nazca.* New York, NY: Memoirs of the American Philosophical Society.

Cotte, M., & Ruggles, C. L. N. (2010). Astronomical heritage in the context of the UNESCO World Heritage Convention: Developing a professional and rational approach. In C. L. N. Ruggles & M. Cotte (Eds.), *Heritage sites of astronomy and archaeoastronomy in the context of the UNESCO World Heritage Convention: A thematic study* (pp. 261–273). Paris: ICOMOS–IAU.

Dakaris, S. (1993). *The Nekromanteion of the Acheron.* Athens: Greek Ministry of Culture Archaeological Receipts Fund.

de Santillana, G., & Von Dechend, E. (1969). *Hamlet's mill.* Boston: Gambit.

Ellis, F. H. (1975). A thousand years of the Pueblo sun–moon–star calendar. In A. Aveni (Ed.), *Archaeoastronomy in Pre-Columbian America* (pp. 59–87). Austin: University of Texas Press.

Fagan, G. (2006). *Archaeological fantasies: How pseudoarchaeology misrepresents the past and misleads the public.* London: Routledge.

Gantenbrink, R. (1999). www.cheops.org.

Hannah, R., & Magli, G. (2011). The role of the sun in the Pantheon design and meaning. *Numen, 58,* 486–513.

Hawkes, J. (1967). God in the machine. *Antiquity, 41,* 174–180.

Hawkins, G. S. (1969). *Ancient lines in the Peruvian desert. Final scientific report for the National Geographic Society Expedition.* Cambridge: Smithsonian Astrophysical Observatory.

Hodder, I. (1999). *The archaeological process: An introduction.* London: Wiley-Blackwell.

Iwaniszewski, S. (2005). Looking through the eyes of ancestors: concepts of the archaeoastronomical record. In J. A. Belmonte & M. P. Zedda (Eds.), *Lights and shadows in cultural astronomy. Proceedings of the SEAC 2005 Conference,* Isili.

Johnson, D., Proulx, D., & Mabee, S. (2002). the correlation between geoglyphs and subterranean water resources in the Río Grande de Nazca drainage. In H. Silverman & W. Isbell (Eds.), *Andean archaeology II: Art, landscape and society* (p. 30). New York: Kluwer Academic/Plenum Publishers.

Kak, S. (2000). Birth and early development of Indian astronomy. In H. Selin (Ed.), *Astronomy across cultures: The history of non-western astronomy* (pp. 303–340). Kluwer.

Kehoe, A., & Kehoe, T. (1973). Cognitive models for archaeological interpretation. *American Antiquity, 38,* 150–154.

Magli, G. (2007). Possible astronomical references in two megalithic buildings of Latium Vetus. *Mediterranean Archaeology and Archaeometry, 7*(2).

Malville, J. M. (2008). *Guide to prehistoric astronomy in the southwest.* Johnson Books.

Malville, J. M., & Munro, A. M. (2011). Cultural identity, continuity, and astronomy in Chaco Canyon. *Archaeoastronomy, 23,* 62–81.

McCluskey, S. C. (1977). *The astronomy of the Hopi Indians.* NY: Science History Publications.

Neugebauer, O. (1975). *A history of ancient mathematical astronomy.* New York, NY: Springer.

Neugebauer, O., & Parker, R. A. (1964). *Egyptian astronomical texts.* Londra: Lund Humphries.

Posnansky, A. (1945). *The cradle of American man.* New York: Augustin Publishers.

Protzen, J. P. (1985). Inca quarrying and stonecutting. *Journal of the Society of Architectural Historians, 44*(2), 161–182.

Protzen, J. P. (1993). *Inca architecture and construction at Ollantaytambo*. Oxford University Press.

Reiche, M. (1980). *Mystery on the desert, Nazca*. Stuttfart: Peru.

Renfrew, C., & Zubrow, E. B. W. (1994). *The ancient mind: Elements of cognitive archaeology*. Cambridge University Press.

Ruggles, C. L. N., & Saunders, N. J. (1993). The study of cultural astronomy. In C. L. N. Ruggles & N. J. Saunders (Eds.), *Astronomies and cultures* (pp. 1–31). Niwot: University Press of Colorado.

Silverman, H., & Proulx, D. (2002). *The Nasca*. New York, NY: Wiley-Blackwell.

Sims, L. (2011). Where is cultural astronomy going? In F. Pimenta, N. Ribeiro, F. Silva, N. Campion, A. Joaquinito, & L. Tirapicos (Eds.), *SEAC 2011 stars and stones: Voyages in archaeoastronomy and cultural astronomy*. London: BAR.

Sofaer, A., & Sinclair, R. M. (1986). Astronomic and related patterning in the architecture of the prehistoric Chaco culture of New Mexico. *Bulletin of the American Astronomical Society, 18* (4), 1044–1045.

Theocaris, P. S., Liritzis, I., Lagios, E., & Sampson, A. (1996). Geophysical prospection, archaeological excavation, and dating in two Hellenic pyramids. *Surveys in Geophysics, 17*(5), 593–618.

Urton, G. (1982). *At the crossroads of the earth and the sky: An andean cosmology*. Austin: University of Texas Press.

Wiseman, J. (1998). Insight: Rethinking the "Halls of Hades". *Archaeology, 51,* 3.

Zeilik, M. (1985). The ethnoastronomy of the historic Pueblos. I: Calendrical sun watching. *Archaeoastronomy, 8,* S1–S24.

Zeilik, M. (1986). The ethnoastronomy of the historic Pueblos: II. Moonwatching. *Archaeoastronomy, 10,* S1–S22.

Part III
Places

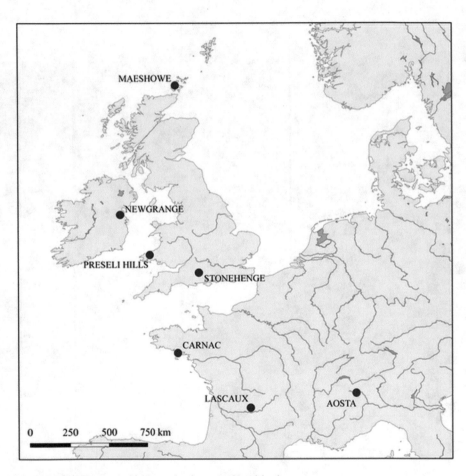

Map of north Europe, with the main sites mentioned in the text

Chapter 7
Megalithic Cultures of the Mediterranean

7.1 Stonehenge and Its Landscape

Astronomy and Stonehenge: almost everyone has heard of this connection, so that Stonehenge has to be the almost obligatory starting point for our archaeoastronomical quest. Stonehenge, however, represents only a small—though, as we shall see, in a sense unique—part of the frenzied building activity that engaged the people of Neolithic Europe and gave rise to the far-flung civilisation that we call "megalithic". The technical skills of these people in quarrying, dressing, moving and erecting huge stones were so refined that, until the development of techniques of absolute dating (in the 1970s), it was often considered impossible for them to have acquired such skills under their steam. As a result, European megalithism was considered as a sort of secondary phenomenon, a knock-on effect of the "diffusionism" of culture emanating from Mesopotamia and Egypt and gradually spreading throughout Europe. It was only with the radiocarbon revolution—the discovery and calibration of the method of absolute dating of organic materials based on radioactive decay—that it became possible to dispel such misguided ideas (Renfrew 1973). Today we know that megalithic architecture in Europe sprang up before or around the fifth millennium BC.

The most simple architectural unit of Megalithic people was the single standing stone, also called the *menhir* (Fig. 7.1). Menhirs are found in isolated positions, or arranged in rows, in some cases numbering hundreds, as is the case in the famous alignments of Carnac, Brittany (Fig. 7.2). It is important to stress that every single menhir is a "constructed" object: it has been quarried, extracted, transported, shaped and erected, in many cases with far from negligible effort and without the use of metal tools. Menhirs were also used to form stone circles, rectangles, and oval-shaped rings.

The second architectural expression of megalithic building is the *dolmen*. A dolmen is a structure consisting of two standing stones supporting a huge lintel, normally (but not necessarily) used for burials. The dolmen principle was also utilised in building complex tombs (which are sometimes, by extension, also called

© Springer Nature Switzerland AG 2020
G. Magli, *Archaeoastronomy*, Undergraduate Lecture Notes in Physics,
https://doi.org/10.1007/978-3-030-45147-9_7

Fig. 7.1 Carnac. The Menhir at Le Manio, height about 6.5 m

Fig. 7.2 Carnac. Alignments of Le Menec

Dolmens) comprising a corridor of orthostats (standing stones) roofed with huge slabs and ending in one or more chambers. In most cases these structures were covered by an earth mound (not always preserved), thus obtaining the so-called tumuli or chambered cairns (Fig. 7.3).

Another typical structure where the menhir is the fundamental unit is the so-called *henge:* a stone circle surrounded by a ditch enclosed in a bank earthwork. And this leads us to the venerable Stonehenge. It is hard to put a label on it, however, as it is architecturally unique. Indeed, in spite of the name and of the presence of a ditch-bank enclosure, it cannot be classified along with the other *henges*. Stonehenge is in fact much more, since the stones of the circle are assembled in a manner which resembles a wooden construction which has been built in stone (Burl 2000). I shall now try to describe this place, conveying both its architectural uniqueness and its (partly true and partly only conjectured) astronomical features.

The monument is located on the Salisbury Plain in Wiltshire, England (Chippindale 1994). It is not set in an especially prominent or privileged position within the landscape. For some reason, the whole area was, however, a place of intense activity for megalithic builders, as testified to by the presence of several other structures of various type, date and function (from tombs to henges, to processional avenues). Stonehenge itself is a remnant of a long process of successive stages of construction, afterthoughts (megaliths were often restructured in various ways), changes and reconstructions that spanned millennia. All these phases have

Fig. 7.3 Carnac. Dolmen (passage grave) of Kermario

been painstakingly classified by archaeologists (only the main points of this classification will be mentioned here); it is worth recalling, though, that the transport and erection of stones can be dated only indirectly through organic material found in relation to them (for instance, inside their foundation holes) so that not all the development phases can be dated with certainty.

The area had been inhabited since very ancient times (8000 BC circa), judging by the discovery of large holes holding huge pine posts. Construction activity underwent a boom in the Neolithic period, around 4000 BC: collective burial sites and the "Cursus" (a large ditch + bank monument, roughly rectangular in shape, 3 km long and 100 m wide) were first created. The first monument at Stonehenge proper (3100 BC circa) was a big circular ditch about 2 m deep and 110 m in diameter, to which corresponded a concentric ring made out of the earth removed from the ditch. Encircling the area, a series of 56 pits—the so-called Aubrey holes —have been recovered, each about a metre in diameter. These pits—used for cremation burials—perhaps originally contained timbers.

The spectacular megalithic evolution of Stonehenge occurred in the phase called Stonehenge 3. During Stonehenge 3I (circa 2600 BC) the builders definitively abandoned timber in favour of stone, and put in place—probably in a double ring and numbering 80—relatively manageable megaliths weighing 4–5 tons each. At least some of these stones, of a quality called bluestone, were transported from as far away as 240 km, from the Preseli Hills, Pembrokeshire, Wales. The fact that the stones came from such a remote area has long been known, indeed it was first suggested in 1923 by the petrologist Herbert Henry Thomas. However, such a feat seemed so utterly incredible—even more so since perfectly decent quarries were available much closer to Stonehenge—that a theory was put forward stating that the bluestones *were already there,* brought by de-glaciation phenomena. Of course, it is extremely odd that no other specimens of bluestone have ever been found dotted around the area, so evidently the builders must have used *every single one available* (for a possible counterexample see Burl 2000). At any rate, every time one comes face to face with a megalithic monument, the best thing is to seek—following petrochemical analysis—the specific place of origin of the stone, that is, not only the geographical area of provenance but the precise quarry (a splendid example is to be seen in Sicily, if one visits the Cusa caves where massive ready-to-go column drums have been patiently waiting to be transported to the Greek Temples of Selinunte for 2400 years). Recently, Neolithic quarries have finally been identified in the Preseli mountains (Darvill 2006; Bevins et al. 2012). The bluestone vein arises naturally in outcrops, such as that on Carn Menyn on top of the Preseli ridge. The cleavage is naturally divided on almost parallel planes, so that the quarrying of rectangular megaliths is relatively easy. Although, as mentioned, they are relatively small (each monolith measures around two metres in height), hauling them from such a far-off place was such a tour de force that naive attempts at replicating the journey today (for instance with the so called "millennium stone project") have failed miserably. Clearly the builders must have had pretty good reasons for dragging the stones from such a remote place; unfortunately, we have little understanding of the significance of the Preseli stone to the Neolithic people, although interesting attempts have been made to identify their interest in the special physical properties of such stone, such as the reflection of light or the emission of sounds when they are subjected to percussion, as well as analysing the cultural (religious) value that the Preseli landscape might have held (Devereux and Wozencroft 2014).

Already during this first megalithic phase the stone ring had a north-eastern entrance (Fig. 7.4). To the same phase probably pertains also the so-called Heel Stone, a monolith located outside the entrance, which originally had a twin stone (now lost) to keep it company. Large portal stones were also set up, of which only the so-called Slaughter Stone remains. Other important features probably relating to this phase were the four so-called Station Stones, located at the corners of an ideal quadrilateral inscribed in the outer circle, the Avenue, a road made out of a parallel pair of ditches which, after an approach set in axis with the monument, bends towards the River Avon, and the huge monolith known as the Altar Stone.

With Stonehenge 3 II (2600 BC–2400 BC) the monument changed drastically. The decision was taken to bring to the site many enormous "sarsen" stones, of a type quarried some 40 km to the north, on the Marlborough Downs. These megaliths were then carefully dressed, for the sections that were to remain above

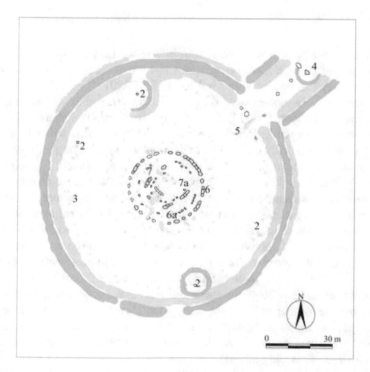

Fig. 7.4 Schematic Map of Stonehenge *1* Ditch-bank enclosure, *2* Station stones, *3* Aubrey holes circle, *4* Heel stone, *5* Altar stone, *6* Sarsen circle, *6a* Bluestones circle, *7* Sarsen Horseshoe, *7a* Bluestone Horseshoe

Fig. 7.5 Stonehenge

ground, while the parts to be buried were hewed with a technique similar to that used for making flint tools, with the curious result that the huge boulders look like macroscopic replicas of Neolithic hand tools (Pitts 2001) (Fig. 7.5).

The sarsens were erected accordingly to quite a complex architectural plan, which included an "outer" ring and an internal structure. The stones of the outer circle were carefully worked in such a way as to be joined together in a circular chain of 30 *trilithons* (trilithon is essentially another way of saying a single dolmen, thus a structure made out of two stone-jambs supporting a lintel). In each trilithon the lintel is fixed with the two door jambs by means of a mortise-tenon joint. The lintels themselves were fitted to each other with a tongue-and-groove joint to form a continuous ring; as far as I am aware, there are no other known examples of this kind of arrangement of megalithic stones. Inside the circular chain, the so-called horseshoe was built. It consists of five *free standing* trilithons arranged in a horseshoe shape; the symmetry axis of the horseshoe is aligned with the north east entrance axis. The stones of these trilithons are even larger than those of the circular chain, and their size and height increase towards the south west with the final ones reaching over seven metres in height. As for the closed ring, the horseshoe is also fairly original in layout and construction (it is sometimes said that that there are analogous examples to be found in megalithic Brittany but these are rather D-shaped rings of standing stones).

What inspired the design of this complex and unique structure? It is difficult to escape the idea that the Stonehenge outer stone circle was meant to suggest a sphere. In fact, not only the *closed* circular arrangement but also the fact that the surfaces of the upright stones widen slightly towards the top, with the lintels curving slightly, recall the idea of a sphere. The interpretation of the horseshoe is difficult, although clearly the idea of some kind of "holy of holies", an inner sanctum, comes to mind; optically, it helps closing the gaps so that the monument looks almost solid when viewed from outside. Approaching Stonehenge from the Avenue, the horseshoe instead appears as the final section of a symbolic path, and the experience is akin to the procession to a temple.

The last stages of the monument (2400 BC–1600 BC) were characterised by somewhat strange activities. First of all, work was carried out to re-erect the bluestones within the sarsen circle. Apparently, however, this adjustment was also considered unsatisfactory, and some bluestones were moved again to form a circle and an oval at the centre of the inner ring. Later on, the north eastern section of the circle was removed, creating a horseshoe of bluestones similar to the central one. Any later associations with the monument are poorly documented in the Iron Age, and thereafter only by a few Roman coins and by some medieval artefacts.

The impression one has from such a long and complex story is that of a sacred place of extraordinary importance. We have no written sources, alas, for megalithic cultures, and to have the first mention of a place which *might* be Stonehenge we must await the Greek writer Diodorus Siculus (first century BC) who mentions a temple of Apollo (the god associated with the sun in Greece) which is "shaped like a sphere" and situated on an island called Hyperborea, which *might* be England. Later legends and speculations about Stonehenge associated it with the Celts, whose culture, however, flourished some 1400 years after the last megalithic phase of the monument.

Precisely because there is no written evidence, any discussion on what Stonehenge was used for or why it was constructed must remain, at least in part, speculative. However, some messages have undoubtedly been written at Stonehenge using the language of stars and stones, and these messages can be read by means of archaeoastronomy.

First of all, the original axis of the monument, as already surmised in the eighteenth century by the antiquarian William Stukeley, points—from the inside looking out—to the rising sun at the summer solstice. The phenomenon is usually photographed today with the sun rising directly from behind the Heel Stone, but the original axis does not pass directly above this stone, but rather was defined by the direction passing between the heel stone and her lost companion. The deliberateness of such an alignment cannot be denied, as we find many other roughly contemporary buildings and tombs exhibiting similar solar alignments; however, the horizon is flat also to the south-west, so the very same axis, when looked from outside the monument, points to the winter solstice sunset (Fig. 7.6).

The spectacle of the sun disappearing amidst the ordered ensemble of giant stones is still quite awe-inspiring today, and it is perfectly possible that—in spite of the modern enthusiasm for the summer solstice alignment—the hierophany was intended as a ceremonial approach to the monument from the north-east and a view —from outside—of the midwinter setting sun *inside* the monument (North 1996;

Fig. 7.6 Stonehenge, winter solstice. The sun sets in alignment with the horseshoe axis (*Courtesy* Clive Ruggles) ⓟ

Ruggles 1999; Sims 2006). Actually, as we shall see, recent analyses of the Stonehenge cultural landscape also point to this alternative interpretation.

Stonehenge is in any case a crystal-clear example of an astronomically oriented building: a building which exhibits a singular and important feature in its design—a symmetry axis in the present case—which is deliberately aligned to a prominent astronomical phenomenon known to the builders and of special interest to them. The source of such interest, as discussed in details in Sect. 5.1, probably resides in the fact that the solar cycle is naturally marked by the two solstices, and usually these 2 days, and in particular the winter solstice, were associated by ancient cultures with concepts of renewal and rebirth.

Were the builders of Stonehenge also interested in the moon's rising/setting positions? At the time of the pioneering studies of alignments at the site by Hawkins (1964), it was proposed that the Aubrey holes were used—according to a somewhat clumsy idea based essentially only on their number, 56—to predict the occurrence of eclipses. As a consequence the monument was dubbed as "Neolithic computer", a debatable label ("abacus" would certainly have been better) which is, nevertheless, still trotted out today.

The curious idea that Stonehenge was used to predict eclipses is discredited by the archaeological evidence regarding the holes; however, other hints point relatively clearly to an interest in the motion of the moon, albeit not at an operative (predicting eclipses) but symbolic level. To understand these indications, let us consider the following experiment. We draw a circle on the ground (in a place with flat horizon), and identify the direction of the sun rising at midsummer/setting at midwinter. Then we put two stones in the circle, in such a way that the cord which joins them lies along this solstitial direction. Now, let us locate the azimuth of the moon rising at the southern maximal standstill/setting at the northern maximal standstill (see Sect. 1.7 for the cycle of the moon at the horizon). We now want to put two further stones on the periphery of the circle, in such a way that the cords joining them to the previous two are parallel to this moon-determined direction, and the figure connecting these points is a rectangle. This is clearly impossible, unless we are at a latitude at which the azimuth of the rising sun at midsummer forms an angle of 90° with the azimuth of the moon rising at the southern standstill. Calculations show that in the northern hemisphere this latitude passes not far from Stonehenge (it is actually in the English Channel). So, the position of Stonehenge is a fairly good marker of this "astronomical geometry", and the monument does indeed have a conspicuous rectangle, the ideal one formed by the station stones, which exhibits these coupled luni-solar alignments. The position of these stones was probably devised before the construction of the sarsen circle, which later obstructed most of the view, and the rectangle was fixed on the basis of the proportion 12:5 so that the sides belong to one of the so-called Pythagorean triplets (triangles with all integers legs, in this case $5^2 + 12^2 = 13^2$) (Ranieri 2002).

So, it would seem quite reasonable to assume premeditation in lunar alignment of the station stones; it cannot even be ruled out that the place of construction of the monument was somehow selected in view of this curious coincidence, although that would imply that several repeated measurements of the two astronomical directions in different places and for several years were made (an example of a sacred place which might have been selected due to astronomical considerations is the Pre-Columbian site of Altavista, see Sect. 7.3).

To conclude, we can certainly say that at Stonehenge there was a clear interest on the part of "megalithic man" in astronomy, as already devised in early studies (Boyle Somerville 1923). However, the way in which astronomy at Stonehenge—and at many other sites—was originally presented by scholars, in particular by Hawkins, stretched this interest too far. How profound really was the preoccupation of megalithic culture with the sky? How accurate were the observations, and to what extent can we speak of a megalithic science? Strongly convinced of the ability of megalithic skywatchers to achieve impressive precision in their observations was one of the founders of archaeoastronomy, Alexander Thom. Thom undertook extensive field surveys of thousands of megalithic sites (Thom 1967, 1971; Thom and Thom 1978; MacKie 1977). According to him, the purpose of many megalithic monuments went far beyond a general interest in the cycles of the sun and the moon, and was centred on the very accurate determination of the days of the solstices and of the positions of the lunar standstills. He further proposed the existence of a megalithic calendar divided into eight seasons and investigated on the use of geometry (in particular, of the Pythagorean triangles, like that of the Station stones) in the layouts of megalithic monuments, suggesting the probable use of a common unit of measure, which he called the megalithic yard. Thom was well aware of the difficulties of high precision naked eye observations, but noticed that these could be overcome by using

Fig. 7.7 Locmariaquier. The Grand Menhir. Today broken into 4 pieces, it once stood for as much as 20 m

a system based on "foresights" and "backsights". The idea was thus that the observers used stones or rows of stones to indicate directions to relevant features (peaks, or notches, or sloping hills) located on the far horizon and marking the rising or setting of the sun or the moon. In this way megalithic people could, at least in principle, fix with great accuracy the solar and lunar cycles.

Most of Thom's claims about the level of accuracy (that he stretched to 1′ in azimuth, which would infer the ability to determine the precise days of the solstices through horizon observations) attained by megalithic astronomy have today been dismissed after comprehensive reassessments of his work (Ruggles 1981, 1984, 1988, 2005). Furthermore, in Thom's days C14 dating was only at its initial stages and now we know that many of Thoms' alignments couple together non-contemporary structures (Fig. 7.7). This is the case, for instance, for the structures which in Thom's view would have made of the Grand Menhir of Locmariaquier—the most massive stone ever erected in northern Europe—a universal foresight for lunar observations; more generally, the whole of Thoms' astronomical interpretation of the Carnac alignments remains quite doubtful. Nonetheless, the bulk of Prof. Thom's work is valid as pioneering groundwork for archaeoastronomy. Personally, I learnt in his work a profound respect for the seriousness of the people who built such extraordinary monuments, and most of all I learnt that archaeoastronomy encompasses the whole landscape in which a monument has been conceived.

Returning to Stonehenge, there is no question that it was *not* a high-precision astronomical instrument, as the alignments between its huge, closely located stones can only define, if anything, very rough astronomical directions. A monument like

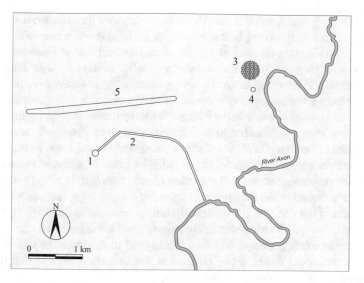

Fig. 7.8 Schematic map of the main Neolithic monuments in the Stonehenge landscape
1 Stonehenge, *2* Stonehenge Avenue, *3* Durrington Walls, *4* Woodhenge, *5* Stonehenge Cursus

this must be understood using the global approach we described in Sect. 6.1, considering it within its own (natural, or man-made) landscape. But then, what exactly was its function? First, it should be said that an artefact with such a long history may have had interchangeable, different and/or complimentary functions. The discovery of the burial sites of individuals who came from distant places, for instance, has suggested that it was a place considered sacred for healing, and the huge efforts clearly invested in its building have led some to suggest that it was a symbol of unity and peace. It is difficult to say; at any rate, it is certainly true that the monument was frequented as a important centre of cult and pilgrimage, and that the surrounding landscape provides us with a clue as to why (Fig. 7.8).

Very close to Stonehenge, at some 3 km to the north-east and near a bend in the river Avon, there once stood two other important monuments The first is Durrington Walls, which was a large settlement enclosed in an oval henge measuring 520 × 450 m. The settlement ringed a number of dwellings, but also some timber circles with no recognisable practical function. It is awaiting complete excavation, but radiocarbon dating of approximately 2600 BC shows that it was contemporary with the main stone phase at Stonehenge. The second structure lies a few dozen metres to the south of Durrington walls, and is nicknamed Woodhenge. It was an impressive construction made up of wooden posts, in the form of a oval ring 85 m in diameter, encompassing six other concentric rings. The complex must have been quite a sight, with posts which, judging by the diameter of the holes, must have been a good 8 m tall and weighed up to several tons: it was thus made of "megadendrs" instead of megaliths (Parker Pearson and Ramilisonina 1998). The positions of the post holes are today marked with concrete cylinders; the result is a little awkward, but allows visitors to have a glimpse at how this site must have been: an artificial forest, furnished however with a main axis—the one with the concentric ovals—which was oriented along the same solstitial line as Stonehenge.

The conspicuous presence of these structures nearby, combined with ethno-historical sources which show that there is a strong link among several native popu-lations (such as Madagascar natives) in the use of the two principal building materials, perishable wood and eternal stone, with the living and the dead respectively, prompted the following idea. Perhaps Woodhenge and Stonehenge were perceived, respec-tively, as "the place of the living" and "the place of ancestors". Stonehenge, located to the west of Woodhenge "where the sun sets", was thus a place of ancestor worship, connected through pilgrimage routes with the other monuments and the river. Further confirmation can be found in the food remains found at Woodhenge, including por-cine, thus associating the site with life and proliferation, whereas at Stonehenge the remains are mostly bovine. As mentioned, Woodhenge has a recognisable axis ori-ented to the summer solstice sunrise, and this would tend to reinforce the idea that Stonehenge was conceived rather to be "reached at sunset", and therefore that the intended orientation was to the winter solstice sunset.

7.2   Newgrange and the Bend of the Boinne

Some fifteen kilometres before reaching its mouth, the Irish River Boyne makes a curious detour, a sort of bend encircling an area some four kilometres wide. This bend—which apparently has no particular geological foundation, it is merely a caprice of the river—is called locally the Bru na Boinne. It is here that one of the most interesting Neolithic landscapes was conceived of and built (O'Kelly 1995; Eogan 1990) (Fig. 7.9).

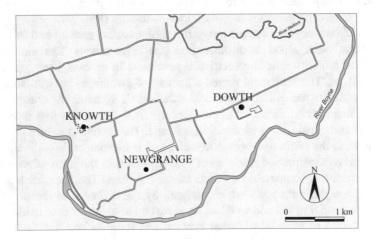

Fig. 7.9 The bend of the Boyne and its main monuments

It is a funerary landscape, a place devoted to the burial and the cult of dead, although, as we shall see, many details of its symbolic meaning remain obscure. In any case, it is difficult to escape the idea that it was precisely the strange course of the river that prompted the choice of this place, all the more so because the area is further enclosed to the north/north-east by a smaller river, the Mattock, so in a sense the bend of the Boyne is naturally isolated by the waters as the henges were isolated by their human-made ditches. Water demarcated here an Isle of the Dead, linked by the River Boyne to the nearby great water, the sea (Lewis-Williams and Pearce 2005).

The Bru na Boinne archaeological area includes three main tumuli, called, going from west to east, Knowth, Newgrange and Dowth, and several smaller tombs, as well as earthworks including a processional way and a circular ditch. These structures are all dated to the end of the fourth millennium BC. Among them, the Knowth mound is the largest (95 m across), and the closest to the river. It is surrounded by eighteen smaller mounds, and equipped with two passages, disconnected from each other and orientated in almost opposing directions. The eastern passage leads to a chamber with three side recesses, covered by means of the so-called corbelling technique, that is, with stone slabs placed in circular layers, with each layer slightly projecting over the previous one up to cover the room. Each of the three recesses contained a hollow stone, or basin, decorated with incisions and used for the remains of the dead. The western passage ends in a single room; the corridor curves to the right about three quarters of the way into the tomb, perhaps indicating the existence of a previous stage of construction which later underwent enlargement. In the passage, there lies a huge entrapped basin.

Newgrange, unlike Knowth, which is circular, has a cardioid shape and is slightly smaller, being approximately 80 m in diameter (Fig. 7.10). The monument is located in prominent position above the Boyne, and dominates the slopes down to the river, as can be clearly perceived as one ascends from the Visitor Centre, which is located close to the river itself. The perimeter of the mound is paved with a kerb of 97 stones and surrounded by a (probably later) stone circle. The quartz-lined façade has been totally reconstructed but (it is claimed by the excavators) is similar to the original one. There is only one passage, ending in a cross-shaped chamber. Decorated stone basins are located in the recesses, and a few human remains have

Fig. 7.10 Newgrange. View from the south of the (reconstructed) mound

been found there, together with burial offerings, like bone beads, pendants and polished stone objects. Similar finds also come from Knowth, the most spectacular one being a Mace Head of carved flint. The axe is covered by stylised motifs which might represent a head, and was certainly a ritual object, to be placed on the top of a post.

Dowth is the easternmost and least well-known of the three, despite being of comparable size. Also here there are two passages leading to burial chambers, but both open towards the south-west and are much shorter than those of Knowth. Thus the chambers do not lie below, or near the centre, of the mound (I wonder if they can refer to pre-existing structures which were inglobated in the monument; in this case, perhaps the main passage has yet to be found, as it possibly occurs for the tumulus which lies unexcavated under the Millmount Fort in the nearby town of Drogheda).

The Bru na Boinne monuments are famous not only for their refined construction techniques but also for their concentration of megalithic art. In fact, many kerb-stones, as well as many stones placed in passages and chambers, are decorated with elaborate motifs—indeed, perhaps only the works of art decorating the slabs of the passage grave of Gavrinis, in Brittany, might be considered superior to them (Fig. 7.11).

The engravings were made by scratching and chipping away at the stones, and many are executed with extreme care and precision. Looking at them the first idea

Fig. 7.11 Knowth. Decorated kerbstones

Fig. 7.12 Knowth. Kerbstone with possible astronomical correlates

that comes to mind is that megalithic art is abstract: it is based on spirals, lozenges, circles, and undulating motives. Only very rarely, as in a Rorschach test, does the observer have the impression of glimpsing at humanoid or animal faces among the motifs, as occurs in the western passage at Knowth, where a stone partially obstructs the way with a carving that evokes a ghostly guardian (yet of course it may be a sort of optical illusion). Ultimately we have no alternative but to define, generically, megalithic art as being symbolic, whatever such a plain definition may mean. Many attempts at interpretation have been made, but we are unlikely to ever receive a definitive explanation. Undoubtedly, some of the decorations hint at relatively familiar astronomical correlates (Fig. 7.12). In particular, Kerbstone 52 at Knowth bears a series of semicircle and circle motifs which have been interpreted as a depictions of successive phases of the moon (Brennan 1994). There are 29 such symbols, a number which is suggestive of a lunar month, but the evidence is not conclusive. Another example are some symbols which resemble the radial lines of a solar meridian; again, it is a hypothesis that is difficult to prove. In any case, most of the decorations do not appear to have been elaborated to convey iconic or factual, but rather spiritual, meaning.

The only way to understand abstract megalithic art remains the interpretation of it within the framework of prehistoric religion, which—as we discussed in Chap. 4 —was most likely based on a shamanistic worldview. Since however the chambers were used for collective burials, the world in question must be the afterworld. The allusions to a realm of the dead to which the chambers give access seem clear: the mounds are kinds of monumental replicas of natural caves, and the waters surrounding the site have the role of symbolic insulation. In what sense the decorations refers to death is difficult to ascertain, but the intermediary with the other world, the shaman, made use of drugs which provoked geometric hallucinations; when represented in drawings, their images are very similar to those of megalithic art. So perhaps the carvings resemble deaths, and/or inform the living about the path that the deceased will encounter, a path already travelled (with a return ticket) by the shaman. Interestingly, some external stones appear to give information about the reliefs present inside; for instance, an elaborately carved motif of three interlinked spirals on the right recess of the Newgrange chamber is replicated on the entrance stone, in front of the monument. Another interesting point is that the decorations included a part of the embellishments which were not meant to be seen, in that they were carved on the reverse face of the stones (the reliefs were only discovered during restoration). The meaning of such hidden art is mysterious, but hints at a possible re-utilisation of pre-carved stones. This practice accordingly may be due to religious, rather than utilitarian, considerations, and is also documented elsewhere. In particular, a astonishing discovery was made a few years ago at Locmariaquier, in the Carnac area. Here, as mentioned in the previous section, the greatest stone ever quarried and erected by megalithic peoples in Europe, the so-called Grand Menhir, was found. It lies today broken up into three pieces, but originally stood more than 20 m tall, weighing some 280 tons. Recent research has shown that it was the tallest of a series of huge menhirs that were erected at the end of the fifth millennium BC. Later, these menhirs were felled, either deliberately or naturally (for example, by earthquakes), but many were re-used in the construction of chambered cairns. In particular, a broken slab from one of them was used for the roof of the huge dolmen called the Table of Marchand. The slab bears an elaborate decoration: an axe, a stick, and the lower part of an ox. The latter etching is incomplete, and it was to great astonishment that, during a restoration of the Gavrinis chambered tomb (located on a island four kilometres away as the crow flies) in 1984, it was discovered that one of the roofing slabs there is indeed the missing piece of this menhir, which was originally 14 m high. The remaining part of the decoration had remained invisible because the slab was mounted upside down, with the decorated part on the other side of the roof (Cassen 2007).

A further point which should be stressed about the Bru na Boinne monuments before discussing the role of astronomy there, is that we need to probe the project choices in greater depth. For instance, the system of corridor and chamber was constructed first, as a separate architectural unit, to be covered later by the corbelling stones and the mound. At Knowth, the basin stone which was located in the Western chamber was (with great personal discomfort, we hope) moved by some thief into the passage, only to realise that the stone was wider than the passage. He then abandoned it. This means that the basin had been prepared and put in place during the first phases of construction, or, as suggested by some scholars, that it was there before the construction of the mound. In this case, the place must already have been considered as sacred, and the mound was later conceived of as a monumentalisation of this central object. Another example of an engineering aspect which probably has a symbolic value is the choice of building materials, as several kinds of different stones were intentionally used in the building works. Many tons of water-smoothed stones were taken from the river bed. Quartz came from the Wicklow Mountains, about 60 km to the south. Granodiorite cobbles probably came from the Mourne Mountains to the north, and Gabbro cobbles came from Dunkald Bay, some 40 km to the north-east. Finally sandstone orthostats were quarried about 10 km to the north. Clearly, all this may be read on two different levels. Firstly, it testifies to a network of social relations and exchange, as well as to the investment of economic capacity in labour for a purely symbolic construction. Secondly, it raises the cognitive question of the symbolic value of the various stones themselves, perhaps meant as the contributions of different communities to the creation of a common place of worship and pilgrimage. In this connection it can be noticed that many stones, besides elaborate carvings, show also recurring, almost childish scores or motifs, which we might perhaps interpret as votive signs left by pilgrims.

The Bru na Boinne was thus a sacred landscape, a place of cultural unity, continuity, and memory, probably used by a class of powerful people to rejuvenate and re-assess the social order in relation to the cosmic order. Clearly, then, we

Fig. 7.13 Newgrange. The (reconstructed) entrance. The roof box is a narrow aperture located at the end of the upper "window" ⏵

would expect astronomy to play a significant role here, and, unsurprisingly, it does. In discussing the archaeoastronomy of the Valley of the Boyne, we cannot but commence from the incident I mentioned in the introduction, which regards Newgrange. It all began in 1963, when, during excavation work, O'Kelly discovered a feature which he called a roof box. It is a stone-built box which acts as a narrow window built over the entrance, with a slight offset. The front edge of the lintel is decorated with a framed series of lozenges (Fig. 7.13).

The window appears to have been equipped with movable slabs used to open and close it at will. This curious carefully designed feature naturally aroused the interest of the excavators. A means for communicating? Or for throwing offerings into the tomb after the closure of the entrance? Possibly. However, it was known through folk memory, passed down over the centuries, that the corridor had something to do with the sun rising at the winter solstice but, because of the upward incline of the passage, the midwinter sunlight could not reach the central chamber, stopping short at some point in the corridor. With the discovery of the roof box, O'Kelly noticed, the situation changed, since the sun could enter from the window and therefore from an increased height, which might compensate for the incline of the passage. This was the idea he attempted to corroborate on 21 December, 1969. He entered the corridor a few minutes before dawn, and saw the sun-rising rays penetrating the corridor *through the window* up to the chamber for the first time after thousands of years. So, this was the original conception of Newgrange: a monument designed from the very beginning to act as a connecting machine between stars (our star) and stones once a year (of course, the movement of the sun at the horizon near the solstices is extremely low, so the light phenomenon actually occurs additionally—though with varying intensities and distributions—on a few days before and after the solstice).

Astronomy is probably also in play at the other two main mounds of the valley. One of the two Dowth corridors has an astronomical orientation to the winter solstice, like Newgrange, but at setting. At Knowth, the two passages are opposite each other and roughly aligned east-west. It is, therefore, tempting to associate Knowth with the equinoxes, as a sort of complimentary monument to solstitial Newgrange. This association, however, has to be taken with a pinch of salt, as recent more accurate measurements have shown that the alignments at Knowth do not correspond precisely to sunrise and sunset at the equinoxes (Prendergast and Ray 2015). The western passage (external straight section) is oriented slightly to the south of west, and the eastern passage slightly to the north of east. Taking into account the altitude of the horizon, it can be seen that the setting sun directly illuminates the western passage approximately seventeen days before the vernal equinox (and likewise seventeen days after the autumnal equinox) while the rising sun directly illuminates the eastern passage approximately six days before the autumnal equinox (and likewise six days after the vernal equinox). Since, contrary to what happens at the solstices, the movement of the sun is rapid at the equinoxes (the daily shift in the azimuth is about one solar diameter) it has been conjectured that it would have been relatively easy for the builders—whose astronomical skills are beyond doubt, by looking at the orientation and the project of Newgrange—to obtain a precise alignment on the precise days of the equinoxes. As a consequence, both here and also more generally, equinoxes are not considered by many scholars to be events of particular interest for the megalithic builders; they are rather considered to be a product of the knowledge of historic, if not modern, society.

In my view, however, it is not so simple. One cannot ignore the importance of cardinality (the division of the world into four equal halves) which follows when north, and therefore the meridian, has been identified. From this, the observation that there exists a day on which the sun rises and sets (with a flat horizon) along the direction orthogonal to the meridian is of course a simple and natural one, and the fact that this day is (approximately) the midpoint between the solstices is also easy to grasp (distinguishing between alignments to equinoxes and to mid-quarters days is a delicate matter, so that interest in the two can be today confused with each other). Furthermore, there are at least a few very convincing alignments to the equinoctial sun to be found in the neolithic period (one of them is the Mnajdra temple in Malta, mentioned in Sect. 4.3). So, I would keep our options open: it may be that Knowth was designed to indicate the equinoxes approximately, or it may be rather that its two passages were planned to indicate quite precisely the *arrival* of the two equinoxes and this, by the way, would explain why there are two passages, and why they are opposite each other.

To sum up, a sort of order seems to be delineated by the astronomical alignments of the main mounds of the Bend of the Boyne. There can be no doubt that although archaeoastronomy is *not* the universal key that dispels all the mist still clouding our understanding of this very special cosmic landscape, astronomy *played an essential part* in its original design.

7.3   The Sleeping Giant

Another region of Europe where megalithic culture produced veritable masterpieces is the Iberian Peninsula. Here we shall concentrate mainly on the megalithic tombs found in the Andalusian provinces of Almeria, Granada and Malaga. These are megalithic sepulchres similar to those unearthed in Ireland: a corridor-chamber structure built of orthostats, and covered by a tumulus. The chamber, at least in the larger exemplars, is trapeziform, with a final wall usually made out of a single, wide backstone and side walls converging towards the entrance. The interior was roofed with horizontal slabs, only a few of which are left, in cases where the tumulus above was not erected or washed away over the centuries. The dimensions of the structures are variable, ranging from just a few metres to the cathedral of megalithic architecture which we shall visit shortly. The tombs are devoid of the elaborate—albeit enigmatic—decorations we encountered in Ireland and Brittany. There is, however, at least one piece of information about the mind of the builders to be gleaned from these huge stones. This information regards the choice they made in the orientation (defined as the direction from the chamber, looking out along the passage) of the entrance passageways (Hoskin et al. 1994). These orientations show a very clear pattern: out of 201 measured azimuths, all but 16 fall in the range 80°–180°. Using the simple approach described in Sect. 3.4, we can estimate the single probability for an azimuth to fall randomly in the interval 80°–180° as $p = 100/360 = 0.28$. The probability of having 185 out of 201 by pure chance in the same interval can now be calculated using binomial distribution and the result is so small as to be considered virtually zero by any test of reasonableness. Therefore, there certainly was a deliberate pattern in the orientation of these megalithic sepulchres. The next step is to prove that the pattern was astronomical and connected to the sun. This is extremely likely since all 185 positive results lie in an interval in which either the sun rises (the azimuth of the sun at the summer solstice is about 60° in this zone, so all azimuths between 80° and 120° correspond to the sun rising at the horizon) or "climbs" on the south-eastern horizon up to culmination (the interval 120°–180°). For the tombs orientated to the rising sun one could further hypothesise that they were orientated on the day on which they started constructing it; tombs orientated to the sun "climbing" were meant to be illuminated fully by the sun each single day of the year, as they still are (no special concentration of data around the equinoxes or the winter solstice occurs).

This example is very instructive, both because it shows a clear-cut case of a deliberate astronomical orientation, proved on a statistical basis, and because, as we shall now see, it reminds us that archaeoastronomy is not only, and cannot function only as, a rigorously exact science. There is, indeed, a outstanding exception to the

Fig. 7.14 Antequera. Entrance to Dolmen de Menga

Fig. 7.15 Antequera. Dolmen de Menga, interior

sun rising/sun climbing orientation rule found in the Iberian peninsula. It is the so-called Dolmen de Menga, in Antequera (Fig. 7.14).

Antequera is a pretty town in the province of Malaga. The town sits on a small hill; approaching it from the east, the enormous stone lintel which covers the entrance of the Dolmen the Menga is perfectly visible (the archaeological park also contains two other monuments: Dolmen de Viera, some 100 m apart, and El Romeral, which lies a few hundred metres downhill). Dolmen de Menga was evidently built by people who were extremely familiar with megaliths. There is no corridor: the plan includes a antechamber, which opens up into a wide chamber formed by seven orthostats per side, 16.5 m long and 6 m wide. The chamber is roofed with enormous stone slabs, supported by three pillars (Fig. 7.15). The monument is a contemporary (if not older) of the passage graves of Ireland, as recent carbon-dating of plant charcoal from pits situated in the atrium has given dates around the middle of the fourth millennium BC (Dolmen de Viera and El Romeral, which each have a much narrower passage, are usually considered to be later buildings).

When visiting the tomb, one cannot fail to notice that the horizon as seen from the tomb is taken up by a rather odd mountain. This mountain, called Peña de los Enamorados, is quite isolated in the landscape and has a peculiar feature: its profile very clearly resembles a gigantic recumbent human head looking up to the sky. This sleeping giant lies only seven kilometres the north-east of Menga; its connection with the tomb becomes dramatic when the visitor first enters the tomb and then

Fig. 7.16 Antequera. The sleeping giant (pena del los Enamorados) as seen from inside Dolmen de Menga

looks back: at this point the human profile fully occupies the view and it becomes obvious that the builders planned the axis of the tomb to obtain this spectacular effect, which we may call a *permanent hierophany*. The mean azimuth of the Dolmen is 45°, while the azimuth of the summer solstice is about 60°, so this special orientation makes the Dolmen de Menga one of the few monuments of Iberia whose orientation is not in the solar range. In other words, direct sunlight does not and did not enter in alignment with the Dolmen—ever—irrespective of the day of the year, although, since the entrance is relatively large, sunlight does hit a part of the right (north) side of the corridor on days close to the summer solstice (Lozanoa et al. 2014) (Fig. 7.16).

The Dolmen de Menga was thus deliberately oriented to the mountain; for some reason, this focus on the mountain was either lost or remained confined to the Menga, as the nearby Viera dolmen is not oriented to it (it is roughly oriented to the rising sun at the equinoxes). Perhaps the whole interpretation of the Dolmen de Menga as a tomb should be revisited, as its uniqueness in orientation and the absence of an entrance corridor rather point to a shrine or a place of worship. Clearly, the mountain is the main candidate for the object of this worship, and this has been confirmed recently by discoveries that show that on La Peña northern side there was a place of ritual activity in the Late Neolithic and Copper Age. This is suggested by the discovery of schematic rock art motifs and of a stratum of microlithic tools spread around a large flat block of white limestone, similar to a natural menhir. The projection of the Menga longitudinal axis passes very close to the location of the rock art shelter (Aguayo and García-Sanjuán 2002).

Many interesting facts emerge at this point. First of all, the Dolmen de Menga is an example of a purely symbolic alignment that is totally unrelated to astronomical observations but can be understood only if the methods of archaeoastronomy are applied to study it. Of course, the Peña de los Enamorados is so glaringly conspicuous to be impossible to ignore, but it is fundamental in *any* survey to have a clear idea of the landscape (natural, or human-built) and of its possible cultural significance. Further, the Dolmen de Menga is a one-off case among the tombs of Iberia. So, in line with the famous Latin saying we encountered in Sect. 6.4, it should be rejected! In exact science, it would certainly be so: in an experiment, a result which is blatantly off the mean would be rejected as spurious. But archaeoastronomy is a multidisciplinary science based on historical remains which cannot be duplicated at will, and this "spurious" Dolmen does in fact have a profound significance.

So whatever our Latin motto says, the fact remains, that the builders of the Dolmen de Menga deliberately oriented one of the masterpieces of megalithic architecture towards one of the most curiously-shaped mountains in the Iberian Peninsula, creating a dramatic and *perennial* hierophany which still can be experienced today, every day.

7.4 Taulas and Stars

The Balearics are a group of Spanish islands located in western Mediterranean. The small (50 km across) Menorca is the easternmost island of the group. The island is nearly flat, apart from its distinguishing peak, the Taurus. As often occurs with island civilisations, a culture with special characteristics developed in this place (and in nearby Mallorca, but the buildings we are dealing with were built exclusively in Menorca). This culture is called Talayotic from the local name for the (hundreds of) huge towers built all over the island. The Talayotic period approximately began around the fourteenth century BC and ended formally with the Roman conquest of the islands in 123 BC, although the Phoenicians established ports much earlier, probably influencing the islanders. The Talayotic culture is characterised by an extraordinary confidence with stone buildings, as witnessed today by a multitude of stunning megalithic monuments and walls of fortified villages, some of them still practically intact—for instance, Son Catlar.

The first of the island's stone-built structures were the above mentioned Talayots, towers erected in dry masonry with large blocks. The real function of the Talayots has never been ascertained, and it may be that the same architectural unit had different functions, in some cases practical (efficient control of the territory) and in other cases and contexts, symbolical. In any case, these iconic towers are scattered throughout the island, and it can be said that their primary function is simply *to exist*. In this respect, they are similar to the so-called Nuraghes, the Bronze Age towers of Sardinia, whose purpose remains obscure, in spite of the established (and ridiculous) archaeological dogma which imputes to the (eight thousand) of them the function of supervising pastures and of chiefs' dwellings.

Fig. 7.17 Menorca. The taula at Torralba

The oblong shape of Talayots, somewhat similar to that of Mount Taurus, which may have inspired it, is repeated in the funerary monument typical of Menorca, the so called Naveta, a megalithic building which owes its name to the fact that it resembles an upturned boat. There are 36 of these monuments surviving, the most famous of which is undoubtedly Es Tudons, a veritable masterpiece of megalithic engineering, 14 m long, which is accessed through a narrow opening in the "stern".

Apart from Talayots and Navetas, megalithic architecture in Menorca is characterised by a third type of building, which is the one of chief interest to us here. If possible, these buildings are even more enigmatic and spectacular than the others. Essentially, they are large ovals of stone slabs stuck in the ground, with a single megalithic monument called *Taula* (the word for table in Spanish) at the centre (Fig. 7.17). Defining a Taula would seem easy: it is an object made up of only two stones, one upright—a sort of column—fixed into a socket excavated in the bedrock, and a second—a sort of capital—set on top of the former. However, such a description does not convey the emotion that these objects can generate. They are, first of all, enormous; secondly, they are very carefully dressed. Finally, they are objects of cult since they prominently occupy the centre of their oval-shaped sanctuaries. It should also be noted that a Taula is not the simplest megalithic construction after the standing stone: it is indeed easier to build a dolmen (lintel over two standing stones), rather than going to the trouble of balancing a huge stone on top of another, and, in actual fact, these monuments developed only in Menorca

Fig. 7.18 Menorca. Taula and Talayot of Trepuco

Fig. 7.19 Menorca. Taula and Talayot of Dalt

and are quite different from all others megalithic structures of the Mediterranean (curiously, by pure chance, they recall the enclosures we encountered at Gobekli Tepe, which were already some 8000 years old when the Taulas were first constructed).

Dozens of Taulas were built: twenty-five are known, but others were probably destroyed in the past (Figs. 7.18 and 7.19). Each Talayotic village had one Taula sanctuary, harmoniously inserted into the village plan. Many of the existing Taulas are well preserved, such as that of Trepucó, whose column stone is four meters high, and that of Torralba, one of the most important villages in the island. They are traditionally dated to the centuries around 1000 BC, and this is the reference date we shall adopt here, backed up also by recent excavations (Sintes Olives and Villalonga 2012). However, it is worth mentioning that, according to some scholars, the construction of the Taulas postdated that of the Talayots by some four centuries and therefore began around the sixth century BC. It is important to warn the reader that this shift in date—at present only a supposition—if proved, would severely jeopardise the astronomical interpretation I am about to present below.

Until a few years ago, some archaeologists—perhaps perplexed by such strange monuments—argued that the Taulas possessed a mere functional purpose, that of supporting a roof. This is clearly a simplistic hypothesis, as the sanctuaries were with all probability open to the sky and therefore the Taulas were the main focus of attention. But which cult are we talking about? Archaeological records are relatively scarce. However, in many of the sanctuaries deposits of animal bones were found, perhaps indicating sacrifices; in Torralba three hooves of bronze belonging to the lower part of a statue of a horse, as well as some Phoenician votive figurines and a statue of a bull, have been uncovered; and finally in Torre den Gaumes a small Egyptian bronze statuette, depicting the god of medicine Imhotep and dated to the seventh century BC, was unearthed. From this complex puzzle one can at least infer that the Taulas were certainly interpreted as places of worship by travellers who came to Menorca, leaving traces of their own religion. As for the Taula cult, the archaeologist Juan Mascaro Pasarius has suggested that the Taulas were stylised images of a bull's head, thus associating them with similar cults operating in the Mediterranean during the Bronze Age (for instance, in Sardinia). Archaeoastronomy, however, can add much to fill this picture in Hoskin (2001).

With one exception, all the Taulas are oriented in a restricted arc of azimuths, all quite close to due south (the horizon is usually flat, as the Taulas either look directly out to sea, or they are on elevated ground and look down over a plain). There is no doubt that this distribution is not random, yet it is too spread out to allow us to think that it just reflected a generic interest in solar illumination of the monuments around midday (a possibility which anyhow cannot be ruled out). Analysing the sky today, we do not find any conspicuous astronomical object at the horizon of Menorca rising in correspondence with these azimuths. But precession has a relevant role if we go back 3000 years in time (Sect. 1.6). Calculations show that the stars of the Southern Cross and of the constellation Centaurus, along with the very bright stars Alpha and Beta Centauri, were visible on the southern horizon in those days. We actually have a pretty good match between the rising azimuths of these stars and the sector of azimuths to which the Taulas are oriented (Fig. 7.20).

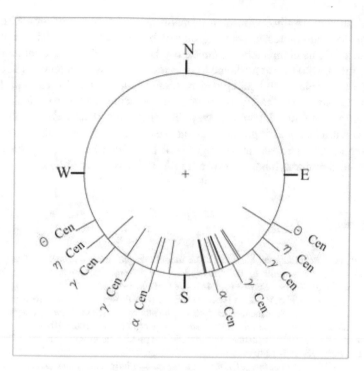

Fig. 7.20 Orientation diagram of the Taulas of Menorca (data by Hoskin 1996)

Fig. 7.21 Menorca. The bright stars of the Cross-Centaurus group as seen in 1000 BC

The Taulas were thus, with high probability, at the centre of a cult associated with the sky and connected with a group of bright stars. The details of this cult, however, elude us completely. In Greek mythology the celestial Centaurus represented Chiron, who taught medicine to the God Asclepius, so a tenuous thread may associate the Taulas with healing and medicine (this would also explain the presence of the statue of the Egyptian god of medicine, Imhotep) (Hoskin 2001). However, we do not know the way in which the Talayotic people viewed Crux-Centaurus stars as a group, and—at least in my opinion—the stars forming our Southern Cross constellation might well have been identified by them with a much more familiar symbol: that of a Taula in the sky (Fig. 7.21).

References

Aguayo, P., & García-Sanjuán, L., (2002) The megalithic phenomenon in Andalusia (Spain): an overview. In *Proceedings of the Colloquium Origin and Development of the Megalithic Phenomenon in Western Europe* (pp. 451–476). Paris: Bougon.

Bevins, R., Ixer, R., Webb, P. C., & Watson, J. S. (2012). Provenancing the rhyolitic and dacitic components of the Stonehenge landscape bluestone lithology: New petrographical and geochemical evidence. *Journal of Archaeological Sciences, 39,* 1005–1019.

Boyle Somerville, T. (1923). Instances of orientation in prehistoric monuments of the British Isles. *Archaeologia, 73,* 193–224.

Brennan, M. (1994). *The stones of time: Calendars, sundials, and stone chambers of ancient Ireland.* London: Inner Tradition.

Burl, A. (2000). *The stone circles of Britain.* New Haven: Yale University Press.

Cassen, S. (2007). Autour de la Table: Explorations archéologiques et discours savants sur des architectures néolithiques à Locmariaquer, Morbihan. Laboratoire de recherches archéologiques, Université de Nantes, Nantes.

Chippindale, C. (1994). *Stonehenge complete.* London: Thames and Hudson.

Darvill, T. (2006). *Stonehenge—The biography of a landscape.* Stroud: History Press.

Devereux, P., & Wozencroft, J. (2014). Stone Age eyes and ears: a visual and acoustic pilot study of Carn Menyn and Environs, Preseli. *Time and Mind, 7,* 47–70.

Eogan, G. (1990). *Knowth and the passage tombs of Ireland.* London: Thames & Hudson.

Hawkins, G. S. (1964). Stonehenge: A neolithic computer. *Nature, 202,* 1258.

Hoskin, M. (2001). *Tombs, temples and their orientations.* Bognor Regis: Ocarina Books.

Hoskin, M., Allan, E., & Gralewski, R. (1994). Studies in Iberian archaeoastronomy: (1) Orientations of the megalithic sepulchres of Almería, Granada and Málaga. *Journal for the History of Astronomy, 25,* S55–S82.

Lewis-Williams, D., & Pearce, D. (2005). *Inside the neolithic mind: Consciousness, cosmos and the realm of the gods London.* London: Thames & Hudson.

Lozanoa, J. A., Ruiz-Puertasb, G., Hódar-Correab, M., Pérez-Valerac, F., & Morgado, A. (2014). Prehistoric engineering and astronomy of the great Menga Dolmen (Málaga, Spain). A geometric and geoarchaeological analysis. *Journal of Archaeological Science, 41,* 759–771.

MacKie, E. (1977). *Science and society in pre-historic Britain.* New York: St. Martin's.

North, J. (1996). *Stonehenge: Neolithic man and the cosmos.* London: Harper and Collins.

O'Kelly, M. (1995). *Newgrange: Archaeology, art and legend.* London: Thames & Hudson.

Parker Pearson, M., & Ramilisonina, (1998). Stonehenge for the ancestors: The stones pass on the messagè. *Antiquity, 72,* 308–326.

Pitts, M. (2001). *Hengeworld.* London: Arrow.

Prendergast, F., & Ray, T. (2015). Alignment of the Western and Eastern passage tombs at Knowth Tomb 1, Appendix 2. In G. Eogan & K. Cleary (Eds.), *Excavations at Knowth 6: The Great Mound at Knowth (Tomb 1) and its passage tomb archaeology*. Dublin: Royal Irish Academy.

Ranieri, M. (2002). Geometry at Stonehenge. *Archaeoastronomy, 17*, 81–87.

Renfrew, C. (1973). *Before civilisation*. London: Cape.

Ruggles, C. L. N. (1981). A critical examination of the megalithic lunar observatories. In C. Ruggles & A. Whittle (Eds.), *Astronomy and Society in Britain During the Period 4000–1500 BC*. BAR British Series 88, Oxford (pp. 153–209).

Ruggles, C. L. N. (1984). Megalithic astronomy: A new archaeological and statistical study of 300 Western Scottish sites. BAR British series 123, British Archaeological Reports, Oxford.

Ruggles, C. L. N. (1988). *Records in stone: Papers In memory of Alexander Thom*. Cambridge: Cambridge University Press.

Ruggles, C. L. N. (1999). *Astronomy in prehistoric Britain and Ireland*. New Haven: Yale University Press.

Ruggles, C. L. N. (2005). *Ancient astronomy: An encyclopedia of cosmologies and myth*. London: ABC-CLIO.

Sims, L. (2006). The solarisation of the moon: Manipulated knowledge at Stonehenge. *Cambridge Archaeological Journal, 16*, 191207.

Sintes Olives, E., & Villalonga, S. (2012). Redescobrint Trepucó: restauració i reinterpretació de la galeria i el barri sud de Margaret Murray. IV Jornades d'Arqueologiade les Illes Balears, Eivissa.

Thom, A. S. (1967). *Megalithic sites in Britain*. Oxford: Clarendon Press.

Thom, A. S. (1971). *Megalithic lunar observatories*. Oxford: Clarendon Press.

Thom, A., & Thom, A. S. (1978). *Megalithic remains in Britain and Brittany*. Oxford: Oxford University Press.

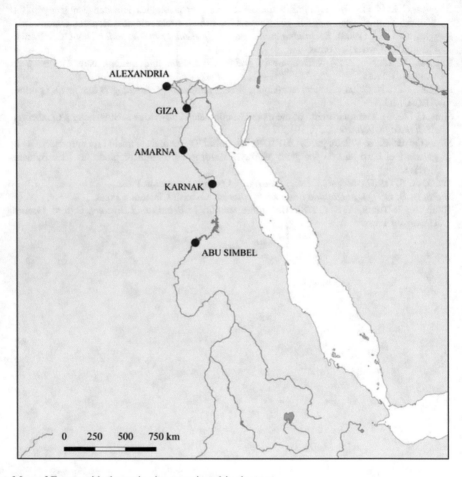

Map of Egypt, with the main sites mentioned in the text

Chapter 8
Ancient Egypt

8.1 A Seat Among the Imperishable

Egypt is a unique place: a short strip of fertile terrain, surrounded by desert, and refreshed only by the River Nile. The climate is consequently arid, but the land is fertile due to the nutritious soil brought by the river during its annual flooding. This land attracted an increasing number of people up until the formation of a unified kingdom about 3200 BC (Shaw 2000; Kemp 2005). The subsequent history of Egypt has traditionally been divided into three "Kingdoms" each distinguished by "dynasties" of rulers; the first period of splendour, the so called Old Kingdom, corresponds to dynasties 3–6, 2630–2152 BC circa (all dates will be given according to the Baines and Malek (1981) chronology).

The King of Egypt, or Pharaoh, invested himself with divine attributes. In particular, a fundamental duty that the Pharaohs assigned to themselves was that of keepers of the cosmic order, called *Maat* by the Egyptians. Maat regulated the world—in particular, the natural cycles and hence the sun and the calendar—and was identified with a goddess, the daughter of the sun God Ra. The ideology associated with kingship thus designated the king as intermediary between gods and humanity, and—as a consequence—the doctrine of power in Egypt was deeply connected with the celestial cycles. Standing witness to this connection we have the so-called Pyramid Texts (hereafter PT), written in the burial chambers of the pyramids—the tombs of the Pharaohs—from the end of the fifth dynasty. These texts are of fundamental importance in understanding not just Egyptian religion but also many key aspects of Egyptian monumental architecture and its relationship with astronomy and landscape.

The PT consist basically of short utterances; they contain, besides rituals, also myths, invocations, magical protections against dangerous beings, and "glorifications", that is, spells aimed at assuring the rebirth of the deceased. The main subject of the glorification texts is the resurrection of the king and his ascent to a celestial afterlife. The deceased ruler, in fact, has several seats reserved in various places on

© Springer Nature Switzerland AG 2020
G. Magli, *Archaeoastronomy*, Undergraduate Lecture Notes in Physics,
https://doi.org/10.1007/978-3-030-45147-9_8

the celestial map. The journey to such places is not easy, given that there are a number of obstacles to be overcome, especially doors, often guarded by sentinels, which the deceased can only pass through by correctly answering certain questions. The topography of the afterworld is thus quite complicated, with multiple directions and corresponding destinations. Let us examine them in some detail.

First of all, there is the region of stars which are defined as *imperishable* and described as follows (all translations of the PT in the present book are taken from Faulkner 1998):

> I will cross to that side on which are the Imperishable Stars, that I may be among them. (PT520, §1223)

In particular, among the imperishable stars there was something called *Meskhetyu*, usually represented as a Bull's Foreleg ⟁. This was the way the Egyptians saw the constellation we call Ursa Major or—more precisely—the asterism of the Plough or Big Dipper, made up of seven brilliant stars. But why was the Big Dipper imperishable? At the latitudes of Egypt in the third millennium BC, this constellation was (with a flat horizon) circumpolar, that is, none of its stars rose or set. Other imperishable stars, besides Meskhetyu, were those of the constellations we call Ursa Minor and Draco. Draco was particularly important because at that time—due to precession—it hosted the north celestial pole. From later documents we know that the ancient Egyptians saw in Draco (or perhaps in a slightly wider area around the pole) a female hippopotamus (Belmonte and Lull 2006). She is usually depicted holding a post, which is mentioned in the PT and probably corresponds to the constellation Bootes and, in particular, to the most brilliant star of the northern sky, Arcturus. As far as Ursa Minor (or rather the Little Dipper, comprising seven stars, like its big sister) is concerned, its stylised form is not dissimilar to that of the Big Dipper, and the Egyptians had a way of indicating the two as celestial adzes.

Interestingly, there is virtually no doubt that these adzes had a terrestrial counterpart in the instruments used by the priests in the *Opening of the Mouth* ritual (Fig. 8.1). This ritual, which has been documented since the time of the Old Kingdom, was then performed on statues of the deceased housed in a specially prepared room, the so-called Serdab (later it would be performed directly on the mummy). The officiant, using special tools, touched the mouth and the eyes of the subject, magically enabling him to receive food, to breathe and to see. The ritual implements used in the ritual are known from both images and grave goods found as burial equipment in priests' tombs; these included, apart from the above-mentioned adzes, a special knife of stone or metal with a forked end. Worth noting in this context is that sometimes these implements are described as being made of iron, a fact which may appear puzzling. However the iron alluded to here is of meteoritic origin: it was available in small quantities because it had fallen to Earth—from the sky—in the form of iron meteorites: a further connection of the ritual with the celestial domain.

Besides the circumpolar stars, a second important area of the sky was that of Sirius and Orion. And indeed the king, as a star, was destined to rise and set there:

Fig. 8.1 The opening of the mouth ceremony from Tutankhamon Tomb in the valley of the kings. The priest Ay ritually opens the mouth of the mummy of the death pharaoh with an adze which has the same shape of the constellation of the Bull's Foreleg (the Big Dipper). In this way, he also asserts his rights in ascending to the throne after the deceased

You will regularly ascend with Orion from the eastern region of the sky, you will regularly descend with Orion into the western region of the sky, your third is Sirius, pure of thrones, it is she who will guide both of you. (PT442, §821–822)

Thus Orion, *Sah* for the Egyptians, was depicted as a celestial dimension of Osiris, God of the Afterworld, and Sirius was identified with his sister Isis.

Finally, but no less important, there is the solar component of rebirth. The king has indeed a reserved seat on the boat of the Sun God Ra:

May you traverse the sky, being united in the darkness; may you rise in the horizon, in the place where it is well with you. (PT 217, §152)

Ra was worshipped in various forms, each associated with a different aspect or role. Ra-Atum was the creator of other deities as well as human beings. But Ra, of course, travelled across the sky, so that another of his guises, the falcon god Ra-Horakhti, was identified with the morning sun; the sun disk itself was considered to be the visible body of Ra. During the night, Ra entered the underworld, only to be reborn the following dawn. Lastly, Ra had a direct influence on the earth, governing the seasons; this influence was recognised especially in the association of the summer solstice with the (approximate) beginning of the gradual flooding of the Nile.

A few other points are worth making with regard to the geography and the inhabitants of the afterworld. In particular, the topography of the PT is completed by a "Winding Waterway" crossing the sky, and by two regions called the "Field of Offerings" and the "Field of Reeds". Clearly, a mirror image of the Nile landscape is perceptible here, and I share the views of many who are convinced that the celestial waterway must be the Milky Way, seen as a celestial counterpart of the Nile as well as one of the personifications of Nut, the sky Goddess. As a matter of

Fig. 8.2 Giza. A view from the east of the Great Pyramid (*foreground*), with Khufu's queens' pyramids on the *left* and Khafra's and Menkaura's pyramids in the *background* 📷 ▶

fact, many civilisations—for instance the Incas—have seen in the Milky Way a celestial river, which could be used as a shaman's path to the afterworld.

These complex religious ideas were essential for the stability of the kingship, since the ruler was identified with a living god, always the same one, Osiris' son Horus, and therefore assuring the afterlife of predecessors was fundamental for the consistency of the entire canonical apparatus. As a consequence, great efforts were put by the state into the construction of "machines" devoted to the preservation of the Pharaoh's bodies and to cults aimed at perpetuating the Pharaoh's afterlife: the pyramids and their annexed temples (Lehner 1999) (Fig. 8.2).

Dozens of royal pyramids were constructed in Egypt, but there can be little doubt about which stand out as masterpieces: the pyramids built by Snefru at Dahshur and those built by his successors Khufu, Khafra and Menkaura at Giza, a desert plateau to the west of modern Cairo (Fig. 8.3). In particular, we find at Giza the unquestionable *chef d'oeuvre* of pyramidal architecture, the Great Pyramid of Khufu. As we shall now see, the ideas on celestial afterlife described in the PT were inserted into the design of this pyramid in such a complex way that it is impossible to understand this marvellous object without the benefit of archaeoastronomy.

The pyramid is built on a rocky spur, abutting directly on the ridges of the plateau, in spite of the fact that the most favourable position would have been that (slightly to the south) of the still-to-come second pyramid of Giza, where the plateau slopes gently, and from a higher point. The monument has a 230.35 m side base (the differences between the lengths of the different sides are less than 10 cm) and was originally about 146 m high; the slope is 14/11 (slopes were measured in terms of rational tangents, that is, using right triangles with integer legs).

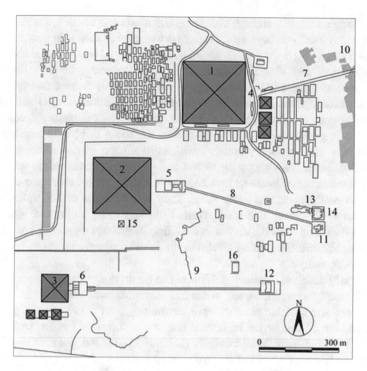

Fig. 8.3 Map of the Giza Necropolis. 1 Pyramid of Khufu. 2 Pyramid of Khafre. 3 Pyramid of Menkaura. 4, 5, 6 Funerary temples. 7, 8, 9 Causeways. 10 Modern village of Nazet. 11 Valley temple of Khafre. 12 Valley temple of Menkaura. 13 Sphinx. 14 Sphinx temple. 15 Satellite pyramid of Khafre. 16 Khentkaues tomb [◪] (▶)

The pyramid consists of a core of 210 horizontal courses of limestone blocks, and was cased with thousands of huge blocks of white limestone (today almost all of them lost). In fact, the result is an artificial mountain, provided, however, with a system of inner rooms, created during the construction (Lehner 1985a). The original entrance (today visitors enter from a passage probably hewn out by looters) is on the north face, and gives access to a descending passageway. Proceeding straight down, this corridor leads to a chamber carved out of the bedrock, which, however, was left rough, or unfinished. Halfway down the descending passageway, an ascending corridor branches off from a gap in the ceiling, but it has been blocked since the time of construction by a plug made of huge blocks of granite. The looters' entrance avoids the plug, and allows us to cross alongside it and begin the ascent to the king's apartment. The ascending passage comes out in the so-called Grand Gallery, a huge room 46.6 m long and 8.54 m high. A recent, non-invasive exploration has shown the presence of a significant void, or chamber, located over this gallery and sealed at the moment of contruction.

At the end of the ascending corridor through the Gallery, a horizontal passage branches off, leading to the so-called Queen's Chamber. This chamber has a projecting corbelled niche on the east wall and two little rectangular openings of shafts, each the size of a handkerchief, visible on the north and south walls respectively; otherwise it is totally anonymous. The niche was clearly meant to house a statue, a fact which firmly supports its interpretation as the place devoted to the ceremony of the Opening of the Mouth. Coming back and ascending through the Gallery, another narrow horizontal passageway is reached. This passes through a small chamber which contained a locking system of granite portcullises. Eventually this chamber takes us to Khufu's burial chamber, or the King's Chamber. Inside there is only Khufu's granite sarcophagus; otherwise the room is bleakly unadorned and silent. It was very probably prefabricated, with all the blocks dressed and their joints tested, before moving up the stones to the course of core blocks where it was actually constructed. Also here, however, as in Queen's Chamber, two small holes are present on the north and south walls. They lead to shafts which, unlike those in the Queen's Chamber, exit the pyramid through the north and south faces respectively. These exits are almost at the same height (about 80 m), and since this is much higher than the starting level, the shafts twist sharply upwards, crossing the pyramid diagonally.

As already discussed in Sect. 6.4, there can be no doubt that the construction of these seemingly straightforward structural elements was actually an extremely sophisticated and intricate piece of work, continuing over many years, together with the completion of successive horizontal courses of the stones of the pyramid. The inclinations of the shafts were carefully maintained over the years, at least, as soon as the most delicate part of the construction in the core of the monument had been successfully carried out. Our task now is to understand why.

The answer is given by archaeoastronomy, and derives from the profound connection between the pyramid and the stars. Yes, there was a intimate connection between pyramids and stars—this should come as no surprise, as the pyramids are the tombs of the Pharaohs, and the deceased Pharaohs were destined to ascend to the sky. The connection of Khufu's tomb with the stars was made effective in two ways, both based on astronomical observations: the incorporation into the monument of four *stellar shafts,* and the orientation of the sides along the cardinal directions. Unfortunately, too much nonsense has been published on the subject "pyramids and stars", so that one can easily be misled: it can become hard to distinguish what is true, what is likely, and what is pseudo-archaeology only fit for the rubbish bin. For this reason I shall proceed with the utmost care.

Let us begin with the shafts of the King's chamber. They have sometimes been interpreted as ventilation channels; this idea is so misguided that I cannot understand how it could have gained so much currency in Egyptological literature. And since the idea still lingers on irritatingly, why don't we spend some time taking it out of the equation for once and for all? Of course, once sealed, the tomb did not need ventilation, so the only possible reason for aerating the chamber—if any, since the channels have been obstructed for centuries but visitors have always accessed the chamber breathing without difficulty—was during the construction. But there are two channels rather than one, and they cross the pyramid diagonally, while for the purposes of ventilation during the construction, building a single vertical shaft would have been a far quicker, simpler and more efficient expedient.

So, if they are not ventilation channels (as they definitely are not), why did the ancient Egyptians built them? The fact that the shafts start from the north and the south sides of the chambers, combined with the fact that the sides of the pyramid are oriented cardinally, means that they are directed to specific points along the celestial meridian. Given that the afterlife was so closely bound up with the sky, as we learnt from the Pyramid Texts, it is natural to look at what was happening at the time of construction at these particular points during the night (Badawy 1964; Trimble 1964). To do this I shall use the data on the shaft slopes deriving from the exploration by Rudolph Gantenbrink (1999). This data refers to the final parts of the channels, where they are reported to be very straight. Hitherto they had to cross the core of the pyramid (where, for instance, the Grand Gallery had to be avoided by the northern channels) and so they have bends and changes of slopes. In any event, the calculations which follow should be taken with a tolerance of, say, plus or minus one degree. Having said that, and fixing as reference the year 2550 BC (a reasonable dating for the reign of Khufu) it can be seen that the north shaft, which is inclined at exit 31° 12′, fits very well with the upper culmination of the "pole star" of the epoch Thuban at 31° 03′, while the southern shaft, which is inclined 45° 00′, points towards the culmination of the central part of the Orion constellation, the so-called Belt, comprising the three stars Al Nitak, Al Nilam and Mintaka, the latter being the closest at 44° 48′ (Fig. 8.4).

The two shafts thus point to the culmination of stars which are mentioned in the PT as main destinations of the Pharaoh's afterworld. A definitive confirmation that this is the correct interpretation comes from the lower shafts. Inclination of the southern lower shaft, after an initial horizontal section, is well attested to as being 39° 36′. This matches the culmination of Sirius at 38° 28′ (Bauval 1993) (the channel actually points to the culmination of also another bright star, Shaula, at 38° 40′). For

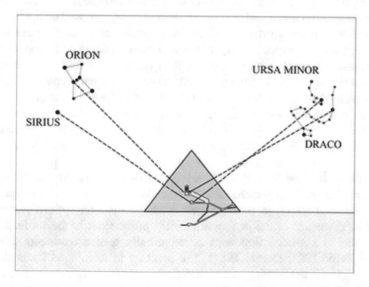

Fig. 8.4 A schematic representation of the astronomical alignments of the four shafts of the pyramid of Khufu. In reality, the shafts start with an horizontal section and exhibit several changes of slopes in their initial run 〽 ⊙

the northern shaft, there is some confusion since the data by Gantenbrink give a value lying between 33 and 40°, but scattered pieces of further information from the (never published) exploration carried out by National Geographic apparently favour the idea that—analogously to the upper ones—the two lower channels were also projected to stop at the same height. Since the chamber is on the vertical of the apex, if this is the case, then the northern lower shaft also has a slope $\sim 39°$ and thus targets well the culmination of Kochab, occurring at 39° 15'; in other words, by a bizarre coincidence Sirius and Kochab had close maximal altitudes—respectively at due south and at due north—at that time.

So, four out of four channels of the Great Pyramid pointed, at the period of construction, to the upper culmination of four stars which play a key role in the Pyramid Texts. There can be no doubt: the function of the channels was to direct the deceased Pharaoh to his proper destinations in the sky.

It is important to stress that the shafts are a purely symbolic architectural feature, as they do *not* "fire like shotguns" to their stellar targets. Moreover, only the upper shafts were projected to exit the pyramid. Why? Up to the beginning of the 1990 s, it was even thought that the lower shafts had simply been left unfinished by the builders, although nobody had ever dared to probe them in depth. The exploration was finally attempted in 1993 by Rudolf Gantenbrink with the help of a small robot, who succeeded in reaching the end of the southern shaft. The robot discovered that the channel had been faultlessly finished using fine limestone blocks, and blocked with a door: a portcullis slab fitted with two copper handles. A similar structure is also shared by the northern channel, as revealed in a second expedition carried out by the National Geographic in September 2002. On this occasion, in the southern shaft a (undocumented) failed attempt at opening the door caused the breakage of one of the handles. Then a drill was applied to the door, and a front-end camera was inserted to see inside, showing a further slab. A later exploration showed that the interior space between the slabs is empty, and that three red ochre signs are traced on the floor. They are probably numbers written in cursive hieroglyphs, and therefore are mason's marks that relate to a "service" part, not meant to be seen, not even symbolically. The doors represent the gates which feature in the Pyramid Texts and which have to be passed through by the soul of the deceased. As such, they would play an important role in making operational the magical workings of the Serdab, that is, the Queen's Chamber from which the channels start. Finally, there is the distinct possibility that the northern channel's door gives (symbolical) access to the chamber discovered over the Grand Gallery; in this case, the chamber might contain—as I have recently proposed—an equipment related to the celestial afterlife and mentioned in the PT (Magli 2017). It is an "iron throne": a wooden chair similar to that belonged to Khufu's mother and known archaeologically, endowed with inlets of meteoritic iron.

As mentioned above, to perform their symbolic function optimally, the upper shafts take advantage of the perfect orientation of the sides to the cardinal points. Clearly, this is a second important astronomically related feature of the Great Pyramid (Petrie 1883; Dorner 1981). The accuracy in orientation attained by the ancient builders is astonishing: the pyramid of Khufu is on average oriented within 3' from the true north.

Fig. 8.5 The Egyptian
astronomer and his assistant
making measures with the
Merkhet

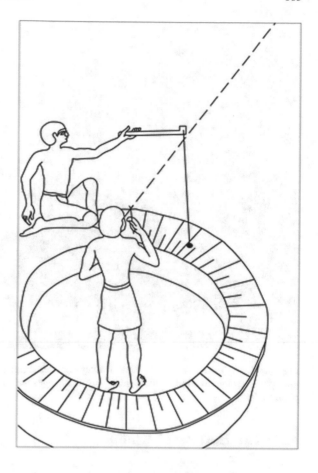

There is no question that accurate star-watching was used to achieve this result,
so that to establish a second, close connection of the monument with the stars. The
problem, however, of exactly how such an accuracy was obtained is quite tricky
and most probably involved a refined method of sighting at certain configurations
of circumpolar stars, called simultaneous transit (Spence 2000; Belmonte 2001;
Magli 2009). In any case the device used by the Egyptian astronomers, called
Merkhet, is known from the archaeological records. It was based—like similar
devices which were to be in use among the Mayas and the Incas—on a fork-shaped
foresight, a palm stem with a slit at the bottom of one side. The instrument was used
as an aid for the naked eye, a sort of viewfinder—of course without any optical aid.
In spite of this, it has been shown that the judicious use of such a device, imple-
mented with a sighting vane, can reach the almost incredible accuracy of ±1′
(Kolbe 2008). The instrument was completed by a level with a plumb line, used by
an astronomer's assistant to project celestial points (for instance, positions of circumpolar stars) at the horizon level (Fig. 8.5).

Fig. 8.6 Giza. A front view of the Khafra valley complex. From *left* to *right* the valley temple and the Sphinx temple in front of the Sphinx. In the *background* the pyramids of Khafra (*left*) and Khufu (*right*)

8.2   The Horizon of Khufu

As we have seen, the Pyramid Texts speak very clearly not only of the stellar destiny of the Pharaoh, but also of a connection of the deceased with the God Ra, and therefore with the Sun. Clearly then, we should expect this connection to be visible at Giza, and indeed, it is so.

Each pyramid complex included two temples located on the east side, looking towards the Nile: a "valley" temple, downhill near the line of maximal flood, and a "funerary" temple located uphill; the two were connected by a monumental causeway. At Giza, the best preserved of such complexes is that of the second pyramid (Fig. 8.6). The two temples here are made of giant limestone blocks, and were once encased with huge slabs of granite. Immediately to the north of the Valley Temple, a pre-existing natural rock was carved as a huge Pharaoh head, and the rock ground was excavated as a lion body, obtaining the Great Sphinx. In front of the Sphinx is another megalithic temple, probably dedicated to it. It is from the terrace in front of this spectacular ensemble of monuments, that one of the most spectacular hierophanies ever devised can be witnessed. To understand how this hierophany works, let us imagine to watch sunset from this terrace, day after day:

the Sun is thus seen to set behind the artificial horizon formed by the pyramids of Giza. In particular, at the winter solstice, the sun sets slightly to the south of the Menkaura pyramid. Approaching the spring equinox, the sun sets behind the second pyramid, up to the day of the equinox when it is seen to set in line with the northern side of it. As we approach the summer solstice, the setting sun will move towards the first pyramid; the northernmost setting point is located in the middle, between the two giant monuments, so that our star is seen setting between the two in the days around the summer solstice (Lehner 1985b). The Sun then, together with the two enormous artificial mountains, forms, on those days, an image which happens to be a spectacular replica of a hieroglyph. This hieroglyph is 🦅—*Akhet*, and represents the sun setting (or rising) between paired, symbolic mountains, the latter being another hieroglyph sign of its own which was called ⌂—*djew*. We are sure that this dramatic phenomenon is an intended hierophany, since the double mountains sign was associated with the afterlife. This association had been made since the Pharaohs of the first two dynasties, who chose for their burial places the desert valley of Umm el Quaab at Abydos. Here, the mouth of a wadi (a dried river) naturally forms the image of a pass between two mountains. Later, the two giant pyramids of Snefru at Dahshur were probably conceived together to form a 🦅 sign when viewed from the Saqqara Necropolis (Magli 2010b, 2011; Belmonte and Magli 2015). Finally, at Giza, the symbol was connected with the sun, creating an even more potent image associated with rebirth (Fig. 8.7). The word Akhet, indeed,

Fig. 8.7 Giza. The sun viewed from the Sphinx terrace sets behind the statue at midsummer evening, with the two main pyramids in the background. Really the huge monument becomes "Horus in the Akhet", the name it had in the New Kingdom, in these moments △ ⊙

when written syllabically (⬭🜚𓇳) 𓅭, has the root—*Akh* ◠ spirit, and in the pyramid texts Akhet is a place where the dead are transformed, preparing themselves for the afterworld. As part of the sky, it was also the place in which the sun, and hence the king, was destined to be reborn. Finally, the choice of the summer solstice for the date is clearly not coincidental, since it heralded the flooding of the Nile.

The question naturally arises—who was the king who devised and ordered the construction of such a majestic spectacle? The solution comes from the fact that we know the *names* of the pyramids of the fourth dynasty. From Snefru onward, indeed, each royal pyramid complex was to receive a name, which in most cases has been passed down to us in the reliefs of the tombs of officials and priests. In particular, we can learn the name of the pyramid of Khufu from many sources, and it was △ 𓏋 ⟐ ⊙, that is, "The Akhet of Khufu" (the name of the king is written first, in the "cartouche"). Therefore, the name passed down to us for the Great Pyramid is *the same* as that of the spectacular hierophany occurring at Giza at the summer solstice. Of course, the fact that the hierophany needs both the two giant pyramids of Giza to be effective raises the issue as to whether the Khufu project already included the plan of the second pyramid. As it happens, there are many indications that this was so (Magli 2003, 2009).

To sum up, the Akhet hierophany replicates in the sky the hieroglyph which was also the symbolic version of the name of the Khufu project. This is, at least in my view, a reflection of a key feature of ancient Egyptian cosmology. In fact, the very act of creation in the Egyptian conception appears to be connected with writing, in a structural analogy between language and cosmos. This analogy is based on a one-to-one relationship: the totality of creation is said to be made of: "all things, all hieroglyphs". A written sign was thus connected with a real thing in a manner somewhat similar to the relationship between thing and concept later elaborated in Greek philosophy (Assmann 2003, 2007). In many cases, a hieroglyph was an idealised but easily recognisable image of an architectural object (a pyramid ⊙, an obelisk ⊙, and so on), but in other cases it was architecture that replicated abstract hieroglyphs. For instance, the hieroglyph ⊙ represents a stylised loaf of bread on a reed mat and means "offering, altar", and in Egypt hundreds of stone altars shaped as giant replicas of this hieroglyph are to be found, the most spectacular being the four-sided altar of the Sun Temple of the fifth dynasty king Niuserra at Abusir. A very complex interaction, therefore, existed between the production of written icons and that of tangible, architectural things: invention in sacred architecture and invention in sacred writing were mutually interchangeable (Fig. 8.8).

This impressive complexity of the Egyptian interplay between creation, writing, and architecture can be better understood in the framework of cultural memory: the preservation of collective knowledge from one generation to the next, and the construction of a collective identity through a shared past (Assmann 2007). Such a cultural continuity, combined with innovation, is a striking leitmotiv running through Ancient Egyptian history. This, in particular, holds for the Akhet symbol.

Fig. 8.8 Abu Gorab. The view towards the 5th dynasty pyramids from the upper terrace of Niuserra's sun temple. A huge alabaster altar is visible in the court

Fig. 8.9 Amarna. The "small" temple of the Aten ⊙

For instance, in the New Kingdom the Sphinx began to be called "Hor-em-akhet" that is "Horus in the Akhet", and the sun between two mountains became synonymous with the Pharaoh's tomb in official documents and texts. Incredible as it may seem, we find it replicated in a spectacular way also in a place which was by other means a truly ground-breaking, revolutionary Egyptian site: Amarna (Fig. 8.9).

Amarna, in Middle Egypt, was the place chosen by Pharaoh Akhenaten (who accessed to the throne around 1350 BC) to establish a new capital. The king was the herald of a brand new religion: a monotheistic cult devoted to the Aten, the Sun Disk (Redford 1987). He abandoned the traditional place of residence of the kings, Thebes (see Sect. 8.4) and decided to found a new town in honour of the Aten in a fairly unattractive, untouched site (perhaps on account of a total solar eclipse occurring there on May 14, 1338 BC, see McMurray 2005; Magli 2013). The city is located on the east bank of the Nile, and the project also included the construction of the Pharaoh tomb, which was placed in a narrow side valley of a wadi (thereafter usually called the Royal Wadi) that dominated the eastern horizon of the town. This choice is one of the many points of rupture with previous tradition introduced by Akhenaten, since the traditional place of the king's burials had always been on the western bank, where the sun sets (in particular, in the New Kingdom it was in the famous Wadi usually called the Valley of the Kings, in western Thebes). In spite of all these innovations, the king elected to revive the Akhet symbol as a key feature of the plan for his town. Indeed, twice a year, around February 19 and October 22,

Fig. 8.10 Amarna, 22 February. The sun rises in the royal wadi re-creating the name of the town, the Akhet of the Aten (*Courtesy* of Marc Gabolde)

looking from the so called "Small Temple" (one of the two main temples of Amarna) it is possible to see Aten, the sun disk, rising within the mouth of the Royal Wadi, forming the hieroglyph Akhet (Gabolde 2004, 2005). To understand the choice of the date, it should be noticed that—in spite of the difficulties in anchoring in a precise way Egyptian dates with Gregorian ones—the date of the town's foundation which is explicitly mentioned in the official Amarna inscriptions almost certainly fell in the third decade of February in the years around 1350 BC, so that Amarna appears to be the first example of a tradition which linked the orientation of a town with a specific date that was considered to be significant for its foundation (a nexus which was to be employed much later with Alexandria, again in Egypt; see Sect. 10.2). At any rate, what is really striking is that, in a perfect analogy with what was devised at Giza 1200 years previously, also here the aim of the spectacular hierophany was to write a name in the cosmic landscape. The Giza hierophany was in fact intended to replicate the name of the Khufu complex: the Akhet of Khufu. The Amarna hierophany was intended to replicate *the name of the new city*. In fact, the new capital was called *Akhetaten,* the Akhet of Aten (Fig. 8.10).

8.3   The Giza Written Landscape

The interplay between writing, architecture and (cosmic) landscape exploited, for instance, at Giza in Khufu's Akhet, was honed to a even higher level of sophistication by the Egyptians, with the encoding of rules—in a sense analogous to writing rules—which governed the evolution of the sacred landscapes of the pyramids' fields.

Everything began, yet again, at Giza. To read what was written in the cosmic landscape of Giza, we shall start from the complex relating to Khafra's son, Menkaura. His pyramid is small compared with the two giants (66 m tall, 105 m wide at the base) but embraces several impressive *tours de force*. The first and foremost is its position. The best place to build a pyramid, considering the two already standing, would have been to the edge of the rocky plateau, more or less on the same meridian as Khafra's funerary temple and therefore several hundred metres to the east of its actual position.

Yet Menkaura's pyramid is located at a great distance, almost lost in the desert. It was not easy to build it there, as this involved a series of additional, tricky technical problems in the transport of the materials, which included thousands of tons of granite to be used for the casing (the backbreaking work was interrupted on the Pharaoh's death and remained unfinished). Menkaura's position on the plateau thus gives us a very clear hint that the topography was *not* determined by functional considerations: the king was *obliged by a rule* to go there. But what rule? Actually, if one observes a map of the monuments on the Giza plateau, the existence of a global topographical axis can easily be seen (Lehner 1985a) (Fig. 8.11).

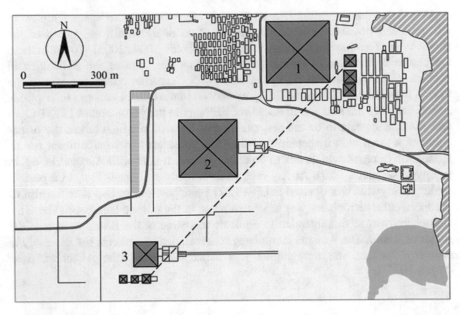

Fig. 8.11 The Giza axis. From *left* to *right* the line runs along the diagonal of Menkaura's first queen's pyramid, the south-east corner of Menkaura's pyramid, the diagonal of his funerary temple, the south-east corner of the second pyramid court, the diagonal of the fore-temple, the south-east corner of Khufu's Pyramid and the diagonal of Khufu's first queen's pyramid

This axis is oriented inter-cardinally (azimuth $\sim 45°$) and is most easily followed by starting from the third Giza complex. The line runs along the diagonal of Menkaura's first Queen's Pyramid, touches the south-east corner of Menkaura's pyramid, follows the diagonal of his funerary temple, passes the south-east corner of the second pyramid court cutting the diagonal of the funerary temple, touches the south-east corner of Khufu's Pyramid and very nearly cuts the diagonal of his first Queen's Pyramid (of course, tracing lines with a pencil or with a computerised ruler on the map of a vast area risks the creation of errors, since the width of the line itself can be of several meters when rescaled on the ground, so it is important to stress that the axis is fairly precise) (Fig. 8.12).

This axis is thus much more than a simple survey line, it is the seminal feature which underpinned the development of the Necropolis. But why? To answer we must adopt a cognitive approach, extending the analysis to the whole cosmic landscape (Lehner 1985b; Jeffreys 1998). The solution lies far to the north-east, on the other bank of the Nile. Indeed, if the Giza axis is projected across the river, it points in the direction of the place where the temple of Heliopolis was located.

Heliopolis was one of the major religious centres of ancient Egypt. It was sacred to the sun and housed the most important temple of Ra; it is mentioned in the PT and was a primary theological centre, where the dominant cosmological doctrine of the Old Kingdom was formulated. This doctrine, sometimes called the Great Ennead because it referred to nine gods, was of tantamount importance because it

Fig. 8.12 Giza. A photograph taken from the first of Menkaura's queens' pyramids, looking north-east. The Giza axis is clearly visible

included the myth of the line of descent of Osiris and therefore underscored the divine nature of the monarch, identified with Osiris' son Horus. Unfortunately, little remains today of the ancient centre, whose ruins lie under the modern buildings of Cairo in a district known as Mataria. The area of the temple is however identifiable by a obelisk (among dozens which originally stood there) erected by Senwostret I, the second Pharaoh of the twelfth dynasty.

In checking that the Giza axis *deliberately* points to Heliopolis, we first observe that the land between the two is occupied by the Nile valley, so in principle an unobstructed line of sight connected the two places in ancient times. However, the earth is round and the distances which come into play are quite significant, to the order of 25 km.

This is clearly too far and would be out of sight for a person looking at the horizon, but by using the horizon formula (Sect. 3.2) and taking into account the fact that that the Khufu pyramid was established on the plateau at about 24 m above the Nile plain, it is simple to confirm that Giza and Heliopolis could communicate (by sun-reflecting signals during the day or fires during the night) using provisional wooden structures, say, no more than 15 m tall (certainly this was not a problem for the builders of stone monuments ten times higher). As a matter of fact, once the pyramid of Khufu attained a sufficient height, the huge monument became visible on the south-western horizon by all visitors to Heliopolis. It was still possible to

Fig. 8.13 Giza. The Menkaura pyramid viewed from the ascending causeway

enjoy this charming view at the end of the nineteenth century, as we can see in paintings of the time.

Thus there is no doubt whatsoever: the Giza axis was deliberately oriented to Heliopolis, and the two places, though quite a distance apart, spoke to each other. To understand why this was considered so essential, we should note that the link between the Sun God and the institution of kingship came to be the most significant element of the ideology of power during the Old Kingdom (Quirke 2001). In particular, the solarisation of the king occurred precisely with Khufu, who to some extent declared himself to be the Sun God. His sons and grandsons consequently defined themselves "Son of Ra" and inserted explicitly the Ra-particle in their names. Since Heliopolis was—at least in a sense—the birthplace of the Sun, an explicit topographical linking of the kings' tombs with the sacred centre becomes perfectly understandable. This relationship can be defined as a *dynastic axis*: a topographical alignment whose symbolic meaning was related to kingship and to the king's blood. It is perhaps also reflected in a passage of the PT (PT 307) where the kings states "My father is an Onite, and I myself am an Onite, born in On when Ra was ruler" (On stands for Heliopolis in the Faulkner translation).

Interestingly, as a consequence of the presence of the Giza axis, anyone approaching Heliopolis and looking towards Giza in ancient times could only see the Great Pyramid, with the other two progressively merging behind it, forming a spectacular optical effect. It is obvious that whenever one puts two pyramids in a restricted area, there will always exist a line of sight along which they are seen to merge. The Giza architects thus intentionally resolved to lay out the second pyramid so that the two merge when seen from Heliopolis, and this was duly followed for Menkaura's as well. So, the Giza axis explains unequivocally why the third pyramid complex was placed so far off in the desert (Fig. 8.13).

One may, however, wonder about the point chosen for the corner of the pyramid along the axis. This point was fixed in such a way that the distance of the apex from the apex of Khafra's roughly equals the distance between the apexes of Khafra and that of Khufu. According to a controversial hypothesis, this leads to a similarity (claimed not to be coincidental) of the three pyramids with the relative position in the sky of the stars of the Orion Belt (Bauval 1989; Bauval and Gilbert 1994). This hypothesis does not hold water: it is in fact clear from the previous section's discussion that the project of the first two pyramids was unrelated to any idea of "replicating" the sky on the ground. There remains the vague possibility that it was Menkaura who made the decision. However, there are other factors which might have influenced the king's choices and can be put forward as likely causes (Magli and Belmonte 2009).

To summarise, then, the Giza axis establishes a direct link between the funerary complex of the kings to the west (Giza) and the "solarisation" of their rebirth to the east (Heliopolis). The inter-cardinal (45°) azimuth of the axis may have had an independent origin, and therefore may have influenced the choice of the Giza plateau. It should in fact be noted that inter-cardinal orientation is very common in sacred buildings in Egypt from the first dynasties on, for instance, in the orientations of the royal tomb at Abydos (Belmonte and Gonzalez Garcia 2009). Its origin might be astronomical, since the inter-cardinal directions south-east/south-west corresponded to the rising-setting points of the most brilliant part of the Milky Way. In particular, during the fourth dynasty, an observer looking towards Giza from Heliopolis would have seen the great celestial river, with the bright stars of the Southern Cross-Centaurus group, disappear from view behind the apex of the Great Pyramid (Magli 2010a).

The end result of all the frenetic planning and building activity which took place at Giza is a stunning example of cosmic landscape, a place where many important details were planned in line with ideas which had more to do with symbolism and religion than with practicality and functionality. This model was to be re-elaborated in the later fifth dynasty, leading to a peculiar structuring of this dynasty's pyramid field, Abusir (Magli 2010a). The evolution of this landscape occurred in fact along a main axis, which is fairly similar in conception to that of Giza, although not functional for a view to Heliopolis. At the end of the Old Kingdom, similar ideas found application also in the pyramid complexes of the sixth dynasty at Saqqara south, where, however, one royal monument, that of king Userkara, is missing and might perhaps be found one day (probably in an unfinished state) in the sands of the desert using archaeoastronomy and Archaeotopography as guides (Magli 2010a, b).

8.4   The Sun in the Temples

After the collapse of the Old Kingdom, Egypt entered a turbulent phase which is usually referred to as the First Intermediate Period. The reunification of the country, with the so-called Middle Kingdom, came about under the Theban kings of the

Fig. 8.14 Karnak. The temple axis at sunset on a summer evening ⊙

eleventh and twelfth dynasties (2134–1783 BC). As a consequence, Thebes (today's Luxor) became the most important city, and the Theban main deity, Amun, progressively rose in importance to the point of being united with Ra, becoming Amun-Ra, the most important Egyptian god.

Together with the advent of the Middle Kingdom, the religious structures and associated priesthood progressively acquired power, and magnificent temples were constructed (Wilkinson 2003). An important part of their foundation ceremony consisted in the choice of the orientation, by means of a procedure called *Stretching of the cord*. In a typical depiction of the ceremony, the Pharaoh is shown together with the goddess associated with wisdom, knowledge and writing, Seshat. The two face each other, and each of them has a hammer in one hand and a pole in the other. Between the two poles there is a rope ring being pulled. The texts associated with the depictions include passages that doubtlessly regard the observation of the northern sky and the circumpolar stars. Indeed, we do find a coherent family of temples orientated along directions close to true north, such as Dendera. The details and the precise targets of such orientations are still, however, the subject of debate (Belmonte et al. 2009).

Another fundamental family of astronomically oriented Egyptian temples is represented by sun-oriented buildings (Belmonte et al. 2009). The most important one is certainly the temple of Amun at Karnak, first founded in the Middle Kingdom (Blyth 2006) (Fig. 8.14).

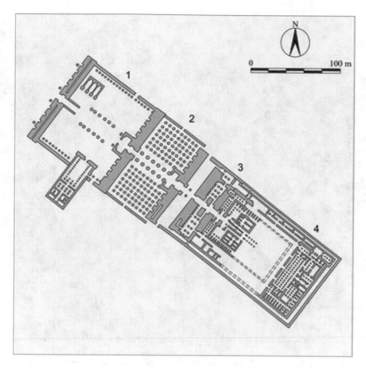

Fig. 8.15 Schematic plan of the Karnak temple proper. 1 First Pylon and court. 2 Hypostyle hall. 3 Hatshepsut's obelisks court. 4 Akhmenu ▶

The axis of the temple runs along the solstitial line midwinter sunrise/ midsummer sunset, with the front towards the Nile and sunset (Lockyer 1894) (Fig. 8.15). However, the western horizon is occupied by the Theban hills, so that the sun sets before filtering along the axis; the horizon is instead flat to the south-east, and therefore the temple is definitely aligned to the winter solstice sunrise: the magnificent spectacle of the rising sun penetrating the temple can still be enjoyed today. But why was Amun-Ra associated with the rising sun at midwinter? This choice probably stemmed from calendrical considerations. In fact, the Egyptian calendar was made up of three seasons. Each season was of 4 months of 30 days, plus 5 epagomenal days for a total of 365 days, without any corrections such as leap years. This has given rise to several—somewhat convoluted—attempts at an explanation (for a complete discussion see Belmonte (2008) and references therein) although, at least in the present author's view, the Egyptians simply knew perfectly well that the calendar was too short to take into account the length of the tropical year, and simply chose to have a calendar comprising the closest whole number of days to those of a solar year for religious reasons. Whatever the truth, the calendar—and thus in particular New Year's Day—drifted along the seasons, making a whole turnaround in about 4 × 365 = 1460 years. Since the calendar started with the summer solstice in the first half of the twenty-seventh century BC,

Fig. 8.16 Western Thebes. The terraces of the Hatshepsut temple at Deir el Bahri

in the years around the foundation of Karnak (say about 2000 BC), New Year's Day completed one half of its "wandering" and was close to the winter solstice (Hawkins 1974; Krupp 1988).

With the passing of the centuries, the Middle Kingdom collapsed and Egypt entered a further turbulent phase which ended again with a reunification, originating at Thebes, which in turn led to the establishment of the New Kingdom (1550–1070 BC). The Pharaohs of the New Kingdom enhanced Karnak with spectacular additions. In particular, this work was undertaken by two very famous rulers: the Pharaoh queen Hatshepsut (1473–1458 BC) and her successor Thutmose III (1458–1425 BC). To legitimise her own rights to the monarchy, Hatshepsut claimed direct descent from Amun-Ra. To validate this ideology, she launched an ambitious building programme on both Theban banks of the Nile. To Karnak, Hatshepsut chose to add a structure open to the south-east. The building was aligned on the same axis as the pre-existing temple but was explicitly linked to the midwinter sunrise, since a group of seated statues representing Hatshepsut and Amun-Ra was the first element to be illuminated by the rising sun, creating a spectacular hierophany. The same orientation was chosen for the masterpiece of the queen's building programme, her magnificent funerary temple located on the western bank in the Deir el Bahri bay (Belmonte and Shaltout 2009) (Fig. 8.16).

On Hatshepsut's death, Thutmose III ascended to the throne and the memory of the Pharaoh-queen swiftly began to fade. In particular, at Karnak the king ordered

Fig. 8.17 Karnak. The temple axis at sunrise at the winter solstice (*Courtesy* Juan Belmonte)

Fig. 8.18   Abu Simbel. The facade of the main temple

the construction of a somewhat unusual structure, the Akhmenu. This was a large, rectangular temple, intended to celebrate the king's ancestors and military campaign. Its longest side was placed in transverse position in relation to the temple axis, clearly showing the intent of obstructing the view of Hatshepsut's buildings along the direction of "her father" rising at the south-east horizon. Interestingly, however, veneration of the rising sun at the winter solstice was reaffirmed with the construction of a side room ("high room of the sun", as Hawkins called it), which is clearly a solar shrine. This elevated building is provided with an oriented window and a huge sun altar. The inscriptions in the room read "We applaud your beautiful face, you biggest of all Gods, Amun-Ra" (Fig. 8.17).

If Karnak is a seminal example of a family of solstitial Egyptian temples, it should be said that solar orientations did not necessarily have to be solstitial. Actually, one of the most impressive hierophanies occurring in Egypt is related to dates in February and October. It occurs at Abu Simbel, where Ramesses II (1279–1213 BC) built two temples excavated in the rock near the bank of the Nile (today reconstructed in an elevated position due to the formation of the artificial lake of the Assuan dam) (Fig. 8.18).

The façade of the main temple houses four gigantic statues, 22 m tall, of the Pharaoh. The interior is organised along a series of halls aligned with one another and decorated with frescos recounting the heroic life of the king. The chapel at the end contains four seated statues representing gods: Ptah, Amun-Ra, Ramesses II as a God, and Ra-Horakhti. On Gregorian dates close to February 20 and October 22 of every year, the sun rises in alignment with the axis of the temple. The light rays pass along the axis and reach the chapel at the very end of the building. However the Sun carefully avoids Ptah (a chthonic God, related to the underworld, therefore remaining in darkness the whole year) but illuminates Amun-Ra, then Ramesses II and, in the end, also Ra-Horakhti, the personification of the solar disk.

Fig. 8.19 Abu Simbel. The innermost chapel of the temple of Ramesses II. From *left* to *right* Ptah, Amun-Ra, Ramesses II, and Ra-Horakhti. Each year, during two short periods of days around February 22 and October 22, from 3200 years, the rising sun aligns with the temple axis and illuminates the statues on the right, while that of Ptah remains forever in the dark ⊙

This spectacular hierophany implies an architectural constraint that conditioned the entire planning of the Abu Simbel main temple right from the onset. To understand the choice of dates we must resort again to the Egyptian calendar. The reign of Ramesses II marks a special calendrical moment in Egyptian history, since for the first time the civil calendar of 365 days re-aligned with the solar (tropical) year, after some 1460 years of wandering. When the calendar was devised—and therefore again during Ramesses II' reign—new years day was the summer solstice and the seasons Shemu and Peret consequently began on Gregorian dates close to those marked by the Abu Simbel alignment, February 20 and October 22 (Belmonte and Shaltout 2005) (Fig. 8.19).

References

Assmann, J. (2003). *The mind of Egypt: History and meaning in the time of the Pharaohs.* NY: Harward University Press.

Assmann, J. (2007). Creation through hieroglyphs: The cosmic grammatology of ancient Egypt. In *Jerusalem studies in religion and culture* (Vol. 6). NY: Brill.

Badawy, A. (1964). The stellar destiny of pharaoh and the so called air shafts in Cheops pyramid *MIOAWB, 10,* 189.

Bauval, R. (1989). A master plan for the three pyramids of Giza based on the three stars of the belt of Orion disc. *Egipt, 13,* 7–18.

Bauval, R. (1993). Cheop's pyramid: A new dating using the latest astronomical data. *Disc. Egypt, 26,* 5.

Bauval, R., & Gilbert, A. (1994). *The Orion mystery.* London: Crown.

Baines, J., & Malek, J. (1981). *The cultural atlas of the World: Ancient Egypt.* Oxford: Oxford University Press.

Belmonte, J. A. (2001). On the orientation of Old Kingdom Egyptian pyramids. *Archaeoastronomy, 26,* S1.

Belmonte, J. A. (2008). The calendar. In J. A. Belmonte & M. Shaltout (Eds.), *In search of cosmic order—Selected essays on Egyptian archaeoastronomy.* Cairo: SCA Press.

Belmonte, J. A., & Shaltout, M. (2005). On the orientation of ancient Egyptian temples: (1) Upper Egypt and lower Nubia. *Journal for the History of Astronomy, 36,* 273–298.

Belmonte, J. A., & Shaltout, M. (2009). *In search of cosmic order—Selected essays on Egyptian archaeoastronomy.* Cairo: Supreme Council of Antiquities Press.

Belmonte, J. A., & Lull, J. (2006). A firmament above thebes: Uncovering the constellations of ancient Egyptians. *Journal for the History of Astronomy, 37,* 373–392.

Belmonte, J.A., & Gonzalez Garcia, C. (2009). The orientation of royal tombs in ancient Egypt. In J. A. Belmonte and M. Shaltout (Eds.), *In search of cosmic order—Selected essays on Egyptian archaeoastronomy.* Cairo: SCA Press.

Belmonte, J. A., & Magli, G. (2015). Astronomy, architecture and symbolism: The global project of Sneferu at Dahshur. *Journal for the History of Astronomy, 46,* 3.

Belmonte, J. A., Shaltout, M., & Fekri, M. (2009a). Astronomy, landscape and symbolism: A study on the orientations of ancient Egyptian temples. In J. A. Belmonte & M. Shaltout (Eds.), *In search of cosmic order, selected essays on Egyptian archaeoastronomy* (pp. 211–282). Cairo: SCA.

Belmonte, J. A., Molinero Polo, M., & Miranda, N. (2009b). Unveiling Seshat: New insights into the stretching of the cord ceremony. In J. A. Belmonte & M. Shaltout (Eds.), *In search of cosmic order—Selected essays on Egyptian archaeoastronomy* (pp. 195–204). Cairo: SCA press.

Blyth, E. (2006). *Karnak: Evolution of a temple*. London: Routledge.

Dorner, J. (1981). *Die Absteckung und astronomische Orientierung ägyptischer Pyramiden*. Ph.D. Thesis, Innsbruck University.

Faulkner, R. (1998). *The ancient Egyptian pyramid texts*. Oxford: Oxford University Press.

Gabolde, M. (2004). The royal necropolis at Tell el-Amarna. *Egyptian Archaeology, 25,* 30–33.

Gabolde, M. (2005). *Akhenaton—Du mystère à la lumière*. Paris: Gallimard.

Gantenbrink, R. (1999). www.cheops.org (Unpublished).

Hawkins, G. S. (1974). Astronomical alignements in Britain, Egypt and Peru philosophical transactions of the royal society of London. *Series A, Mathematical and Physical Sciences, 276* (1257), 157–167.

Jeffreys, D. G. (1998) The topography of Heliopolis and Memphis: Some cognitive aspects. In H. Guksch & D. Polz (Eds.), *Stationen: Beitrage zur Kulturgeschichte Agyptens. Rainer Stadelmann Gewimdet* (pp. 63–71). Mainz: von Zabern.

Kemp, B. J. (2005). *Ancient Egypt: Anatomy of a vivilization*. NY: Routledge.

Kolbe, G. (2008). A test of the simultaneous transit method. *Journal for the History of Astronomy, 39,* 515–517.

Krupp, E. C. (1988). The light in the temples. In C. L. N. Ruggles (Ed.), *Records in stone: Papers in memory of Alexander Thom*. Cambridge: Cambridge University Press.

Lehner, M. (1985a). A contextual approach to the Giza pyramids. *Archiv fur Orientforschung, 31,* 136–158.

Lehner, M. (1985b). The development of the Giza Necropolis: The Khufu project. In *Mitteilungen des Deutschen Archaologischen Instituts Abteilung Kairo* (Vol. 41).

Lehner, M. (1999). *The complete pyramids*. London: Thames and Hudson.

Lockyer, N. (1894). *The dawn of astronomy*.

Magli, G. (2003). On the astronomical orientation of the IV dynasty Egyptian pyramids and the dating of the second Giza pyramid. http://arxiv.org/abs/physics/0307100.

Magli, G. (2009). Akhet Khufu: Archaeo-astronomical hints at a common project of the two main pyramids of Giza. *Egypt Nexus Network Journal—Architecture and Mathematics, 11,* 35–50.

Magli, G. (2010a). Astronomy, topography and dynastic history in the alignments of the pyramids' fields of the Old Kingdom. *Mediterranean Archaeology and Archaeometry 10,* 59–74.

Magli, G. (2010b). Archaeoastronomy and archaeo-topography as tools in the search for a missing Egyptian pyramid. *PalArch's Journal of Archaeology of Egypt/Egyptology, 7*(5).

Magli, G. (2011). From Abydos to the valley of the kings and Amarna: The conception of royal funerary landscapes in the New Kingdom. *Mediterranean Archaeology and Archaeometry, 11,* 2–21.

Magli, G. (2013). *Architecture, astronomy and sacred landscape in ancient Egypt*. Cambridge: Cambridge University Press.

Magli, G. (2017). A possible explanation of the void discovered in the pyramid of Khufu on the basis of the pyramid texts. Preprint arXiv 1711.04617.

Magli, G., & Belmonte, J. (2009). Pyramids and stars: Facts, conjectures and starry tales. In J. Belmonte & M. Shaltout (Eds.), *In search of cosmic order—Selected essays on Egyptian archaeoastronomy*. Cairo: Supreme Council of Antiquities Press.

McMurray, W. (2005). Dating the Amarna period in Egypt: Did a solar eclipse inspire Akhenaten? www.egiptomania.com/EEF/EEFLibrary4.html.

Petrie, F. (1883). *The pyramids and temples of Gizeh*. London: Field & Tuer.

Quirke, S. (2001). *The cult of Ra*. London: Thames and Hudson.

Shaw, I. (2000). *The Oxford history of ancient Egypt*. Oxford: Oxford University Press.

Spence, K. (2000). Ancient Egyptian chronology and the astronomical orientation of pyramids. *Nature, 408,* 320.

Trimble, V. (1964). Astronomical investigations concerning the so called air shafts of Cheops pyramid. *MIOAWB 10,* 183.

Wilkinson, R. H. (2003). *The complete temples of ancient Egypt*. Thames and Hudson.

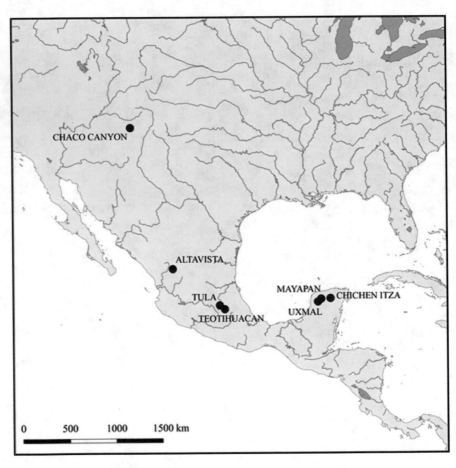

Map of North and central America, with the main sites mentioned in the text

Chapter 9
Pre-Columbian Cultures

9.1 The Maya at Uxmal: The Governor's Palace

The Maya culture flourished in the area which extends from the Yucatan peninsula to south-east Mexico, Belize, Guatemala and Honduras (Coe 2001). Between the third and the ninth centuries AD—the so called Classic period—hundreds of Maya city states developed here; most of them then rapidly collapsed, for unclear reasons. However, in the subsequent or Post-Classic period, equally wonderful cities thrived in the Yucatan lowlands, and we shall concentrate here on two such places, Uxmal (this section) and Chichen Itza' (next section).

Maya city-states were a highly complex construct. They were autonomous, but linked by structured, mutual alliances and inter-dependences, as well as by fierce feuds which often flared up into conflicts. The political dependence between cities was in many cases celebrated with the construction of a *sacbe*. Sacbes are ceremonial straight roads, running on an artificial platform. Some of them are very long, and the longest known—connecting Coba' to Yaxuna'—extend to 105 km, proceeding with impressive straightness, close to the east-west direction along many stretches.

The political history of the Maya cities has gradually been revealed by Mayanists in the last few decades, thanks to the progressive deciphering of inscriptions. The history of such deciphering is very instructive; to put it briefly, it has long been known how the Mayas wrote numbers, so it was also common knowledge that the vast majority of inscriptions contained dates; however, up to the 1950s, a rather iniquitous idea prevailed, namely that the Mayas simply recorded the passing of time. In other words, there was no inkling that the inscriptions could be what they naturally seem to be, and actually are: official recordings of historical facts, for example, the accession of a ruler or celebration of a victory.

The Maya culture and religion were extremely complex. Their tiered cosmos was structured in three levels. As mentioned in Sect. 5.2, the underground world, or Xibalbà, had an intricate geography and was divided into nine sub-levels, each

© Springer Nature Switzerland AG 2020
G. Magli, *Archaeoastronomy*, Undergraduate Lecture Notes in Physics,
https://doi.org/10.1007/978-3-030-45147-9_9

inhabited by a divinity associated with death. The heavenly world was populated mainly by gods associated with natural phenomena, for example Chac, the Rain God. In the sky, Venus was associated with Chac and played a important role—so that her cycles were studied scrupulously, as we shall see shortly. More generally speaking, the prediction of all cosmic events and the study of their presumed influence on human lives was undoubtedly of fundamental significance for the Mayas (Schele et al. 1995). In particular, they developed sophisticated tables for studying the incidence of eclipses, as testified to by the few Maya books—called codices—that have come down to us. Of course, astronomy and astrology were inextricably linked in these texts, since celestial events were connected with rituals and omens.

It has often been said that the Mayas were obsessed by the calendar, and many ridiculous attempts were made to stir up a similar obsession in us all on the occasion of December 2012 (see below). In reality, the Maya were "obsessed" only insofar as calendrics was an essential ingredient of the *timing* of Maya agricultural activities as well as of their social life, from the choice of the name of a newborn to the festivities connected with the completion of time cycles. These cycles were measured by means of three different calendars: the *Tzolkin* of 260 days (13 months of 20 days), the *Haab*, of 365 days, and finally a long-term calendar or *long count*.

The reason for the choice of 260 days for the Tzolkin is unknown. There are actually two astronomical periods which (more or less exactly) last 260 days. The first is the interval between two zenith passages of the sun at a certain latitude, and in the Maya area this latitude corresponds to the classic city of Copan. As a consequence, Copan has been repeatedly proposed as the place where this calendar was devised. The other astronomical possibility relates to the Venus cycle, since the period of visibility of Venus as the Morning Star is about 258 days. In both cases, however, it must be clarified that an astronomically anchored calendar *cannot* be run on the basis of these celestial periods, because both are only *parts* of a complete cycle. Since this point is often misunderstood, I shall explain it with an example. Suppose that we start a calendar of 260 days on the second passage of the sun at the Zenith in a place where the two passages are so separate. Then, the first calendar count will end with the first zenith passage, and therefore it will be perfectly in tune with a celestial cycle. That is all, basically, since with the second period of 260 days, the calendar will already be completely out of sync. Therefore, if the Tzolkin calendar also had astronomical origins, (as yet unproven) the calendar length of 260 days was soon to be left as a purely symbolic fossil of such an origin.

The Haab calendar consisted of 18 months of 20 days each, plus five days considered to be inauspicious, making a total of 365 days. There was no leap year, although its solar origin appears obvious (the Maya were perfectly aware of the true length of the solar year). Finally, dates of the long count were expressed with five figures in this way: kin (day), unial (20 days), tun (18 unial = 360 days), katun (20 tun = 7200 days), baktun (20 katun = 144,000 days), exactly as our dates are made up of days, months and years. For us, though, the days go from 1 to at most 31 and then start again, the months go from 1 to 12 and then start again, but the years go on ad infinitum. For the Mayas, *all* the five figures were recursive: kin, tun and katun went from 0 to 19, the unial went from 0 to 17 and the baktun from 1 to 13. Thus, all possible long count dates, consisting of five numbers in succession, were finite in number. A complete long count period ended on 13.0.0.0.0., after 5125 solar years from the beginning (hence—curiously—five such periods are the equivalent of 26,625 years, a number of years extremely close to a precessional cycle).

Of course, to collocate a Maya date that we read in inscriptions in a moment of our own history we must first *anchor* at least one Maya date to a Gregorian one. Fortunately, this problem is solved by the so-called GMT (Goodman-Martinez-Thompson) correlation, which fixes (on the basis of historical documents reporting astronomical events) the starting date of the long count on August 13, 3114 BC (there is a uncertainty of one day). It can be inferred from this date that the end of the 13 Baktun and therefore the re-start of the Maya calendar was at the winter solstice 2012. However, the catastrophe theories linked to this (besides proving to be downright and embarrassingly false on the day after…) find no justification in the extant Maya documents and records. Obviously, however, a very interesting issue remains on the table and—as far as I know—it is totally unaddressed: that of understanding the reasons behind the choice of the starting date, as well as the length of the recursive period. To make a comparison, an archaeologist studying our remains would have to understand why the Christian world counted time from 1 *Anno Domini* and the Islamic world from 1 *Hijra*.

The role of the astronomers in the Maya society was quite a considerable one, and we do have clear evidence that Maya astronomers were involved in the planning of monumental buildings. The importance of the celestial cycles was so rooted in Maya life that astronomy will often crop up in the urban fabric of cities in the most unexpected ways and forms (Aveni and Hartung 1986; Ashmore 1991). As a consequence, it may be extremely difficult for us to decipher today; furthermore, the new-age mumbo-jumbo which accompanies any discourse on Maya spirituality and cosmology does not help in identifying their true features. In any case, it is beyond question that the orientations in Mesoamerica exhibit a non-uniform distribution, and this is undoubtedly of astronomical origin. This origin is very likely solar. Peaks can be seen at the solstices and at the days (very close to the equinoxes) marking quarter days of the year, i.e. midpoints in time between the solstices (Sprajc 1993; Sprajc and Sanchez Nava 2012); moreover, the intervals separating sunrise and sunset dates recorded by orientations tend to be multiples of 13 or of 20 days. The urban layouts tend to be skewed clockwise from cardinal directions by an amount close to 17°, probably indicating (rather than a skewed orientation of the main north-south axis with respect to the north) an interest in the sun rising in autumn and/or setting in spring. This tradition was perhaps initiated, or at any rate made popular, by the urban layout of Teotihuacan (Sect. 5.5). The true origin of it, however, is still subject to debate.

Besides solar orientations, Maya architecture also presents convincing cases of orientation towards the extreme declinations of Venus. The reason might have been partly practical, since during the Maya period the extremes of Venus had a certain relationship with the rainy season and thus with the maize cultivation cycle. However, this correlation was too approximate to be considered effective for timing agricultural activities. Rather than, the connection was mainly symbolic, as Venus was connected with rain, maize, and fertility in the Maya religion (Milbrath 1999).

The most convincing case of an alignment to Venus is to be found in the Maya city of Uxmal, in south-western Yucatan (Fig. 9.1). Uxmal, together with the closely related towns of the so-called Puuc Route, Kabáh, Labná and Sayil, reached its apex in the Post-Classic period, roughly between AD 700 and 1000. The architectural style which characterises these towns is unmistakable. First of all, a Puuc palace façade is divided into a lower part, with several doorways, and an upper part without windows and decorated with symbolic motifs. The motifs were carved in stone and assembled with a technique which recalls a modern Lego construction. Usually, there are a few single personages, while the rest is occupied by a geometric decoration and by the ubiquitous, obsessive repetition of a peculiar image. The latter is a rectangular mask, representing a stylised face with a very elongated nose (or, according to some, upper lip). This mask is replicated scores, or even hundreds, of times. Each mask is made up of smaller stone pieces, one for each detail (e.g. eyes, nose, etc.) and each of these stone pieces was produced in series (so that in front of the palace of Kabah, which is currently under restoration, long lines of identical noses, identical eyes and so on are to be seen on the ground). Traditionally, the mask is identified with an image of Chac, the rain god, an

Fig. 9.1 Uxmal. General view

identification which seems quite sound (for an alternate view as a "flower mountain" symbol, see Coe 2001).

The Puuc style at Uxmal reached its pinnacle in three main structures. First of
all, the so-called Pyramid of the Magician (Fig. 9.2). It is an unusual building, with
rounded slopes and an oval plan. Second, and connected with the first, the
quadrilateral court traditionally called the Nunnery quadrangle. Both of these
structures follow the general orientation of the whole ceremonial centre, which—as
is usual in Mesoamerica—is slightly east of north. The third building, however, is
orientated 28° south of east, and its façade looks directly onto the jungle (Fig. 9.3).
It is the so-called Palace of the Governor, a huge rectangular building located on a
platform rising up to 14 m above ground level. As is typical in this kind of official
architecture, it is an object meant to be seen rather than to be inhabited, as it is
equipped with as many as eleven doorways on the main façade and has no windows. Its function was therefore mainly symbolic, as a sort of king's official
mansion. The decoration of the palace is in the sumptuous Puuc style, with a
succession of Chac masks. Maya numerals representing the number eight are
sculpted over the eyes of the masks placed at both northern corners of the building,
and hundreds of so-called Venus glyphs are also present on the façade. The Venus
glyph is a symbol which might be described as two stylised eyes underlined or
overlined by moustaches; such a symbol is unequivocally associated with the planet
Venus in Maya iconography. Aligned with the palace's front, are a huge cylindrical

Fig. 9.2 Uxmal. The pyramid of the Magician

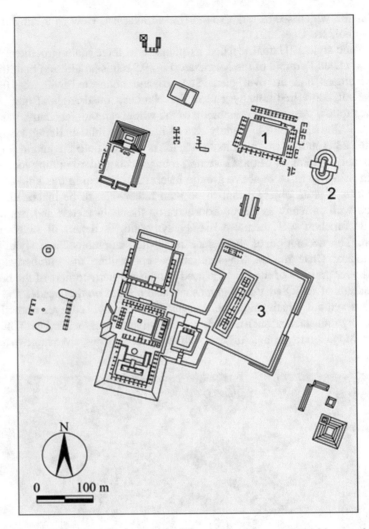

Fig. 9.3 Map of the ceremonial center at Uxmal, with the monuments cited in text. *1* Nunnery quadrangle, *2* Pyramid of the Magician, *3* Palace of the Governor

"menhir" and a square platform with a stone double-headed jaguar throne (Fig. 9.4). The dating of the palace is to around 900 AD, during the reign of a king commonly known as Lord Chac.

The combination of a skewed orientation in relation to the urban plan and the Venus glyphs strongly hint at an astronomical connection, and indeed, the perpendicular to the façade points to the southerly extreme of the rising of Venus in its complete eight-year cycle (Sect. 1.8). Viewed from the Palace, this direction is marked on the horizon by a pyramid belonging to a small Maya settlement located at about 5 km from Uxmal (Aveni and Hartung 1978). Of course, the Palace can

Fig. 9.4 Uxmal. Palace of the Governor

also be used as a foresight for observing the northerly extreme of the setting of Venus from the summit of this pyramid, and perhaps, in view of the behaviour of the extrema of Venus in the centuries in which the palace was constructed, this alternative event is the one which was actually observed (Sprajc 1993). In any case, the connection of the Palace of the Governor with Venus and its consequent deliberate orientation appears unquestionable; the number eight in the decoration of the building may refer to the period of invisibility of the planet at the inferior conjunction and/or to the fact that five synodic Venus periods of 584 days (a periodicity very well known to the Maya) equal eight Haab years of 365 days.

9.2 The Serpent on the Pyramid: Chichen Itza'

The Yucatan peninsula is dotted with peculiar natural features called *cenotes*. A cenote is a sinkhole, in many cases circular or nearly circular, which results from the collapse of the bedrock surface and can be quite massive (up to several dozen metres in diameter). The collapse exposes pits with galleries of groundwater beneath, so that many cenotes resemble natural swimming pools. The cenotes were considered sacred by the Mayas, who used them as gateways of spiritual communication with the other world and, as a consequence, cast sacrificial offerings, including humans, into them.

It is probably on account of the presence of two imposing cenotes that Chichen-Itza, one of the most important Maya sites of Yucatan, was founded. The first settlement here dates as far back as the Classic Period, with the construction of a ceremonial centre which today bears the name of Chichen Vejo. But Chichen-Itza was to become one of the most important city of the Yucatan only in the Post-Classic era, say, around the tenth century AD (Fig. 9.5). Up to a few years ago, it was believed that Chichen Itza' had been invaded and re-founded by the Toltecs, a warrior tribe from the Mexican plateau who allegedly conquered the town

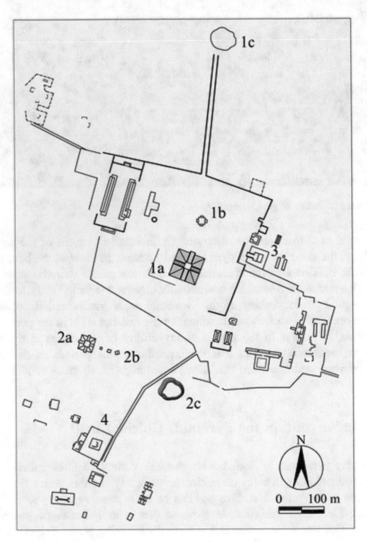

Fig. 9.5 Map of the ceremonial center at Chichen Itza, with the monuments cited in text. *1a* Castillo, *1b* Platform of Venus, *1c* Sacred Cenote, *2a* Osario Pyramid, *2b* 2nd Venus Platform, *2c* Cenote Xtoloc, *3* Temple of the warriors, *4* Caracol

around 967 AD. Recently, doubt has been cast on this invasion (Kowalski 2012). Be that as it may, there is *no* doubting the fact that at Tula, the heartland of the Toltecs in central Mexico, there is a building—the so called Temple of the Morning Star—which has been *copied*, in greater dimensions, at Chichen Itza' in the so called Temple of the Warriors (both having the same, familiar orientation with a 17° skew).

Chichen Itza is one of the most popular and awe-inspiring archaeological sites in Mexico, and a complete description of it would be beyond the scope of this book; we shall concentrate here on the main pyramid called Castillo. However, we must first mention a building, the so-called Caracol, which has provoked much discussion in view of its possible astronomical function (Fig. 9.6).

The Caracol is an enormous structure with circular layout, built over a huge platform. From the outside it admittedly resembles a modern astronomical observatory covered by a domed vault, but the dome is actually almost filled in, apart from a spiral passage that gives the building its name (Caracol means snail in Spanish). Thus, the Caracol's resemblance to modern astronomical observatories is purely coincidental. However, there are many astronomical alignments (to Venus standstills, to solstices, and others) that appear to correspond to particular features of the edifice (Aveni et al. 1975; Aveni 2001). The validity of these alignments has been questioned, though, especially since the corresponding sights from the windows of the upper dome are not direct but are based on cross-jamb angles, and as such are more at risk of a selection effect (Schaefer 2006). Recently, I was fortunate

Fig. 9.6 Chichen Itza. The Caracol

enough to be granted a study visit to the interior, which is generally kept closed for security reasons by the INAH authorities. After this visit I should say that a re-analysis of this building would certainly be worthwhile. However, if many details might be open to question, it is frankly difficult to understand the many curious features—like the apparently odd changes of axes—without having recourse to astronomy. For instance, the stylobate on the front is skewed in such a way to be convincingly oriented to the southernmost extreme of the rising of Venus.

From the Caracol we can now move 400 m to the north and reach the world-famous pyramid usually called Castillo (from the Spanish word for castle). It is a square building 24 m tall, with 55 m side base. Each side is fitted with a staircase inclined at 45° which reaches the summit, which is occupied by a rectangular temple. The monument is dedicated to Kukulcan, the Plumed Serpent, a very ancient divinity in Mesoamerican religion who was, however, also identified (at least mythologically) with a historical figure—some kind of ruler of the town. Two monolithic sculptures of serpent's heads are placed on each side of only one of the staircases; clearly this is the most important one as it leads to the main entrance to the edifice on the summit (Fig. 9.7). In the 1930s it was discovered that the building incorporates, like a Matrioska doll, an older built-in pyramid, with its summit temple still intact, containing a Chac-Mool (a statue of a recumbent man which was used for human sacrifice) and a jaguar-shaped throne. This older structure was oriented like the Castillo but probably had only a single stairway, which corresponds to the main one of the Castillo itself (today it runs underneath it).

In the design of the Castillo we find many numbers recurring, which may allude to calendrical meaning—in particular, each staircase counts 91 stairs, for a total of 365 if the upper terrace is included—although identifying numbers in the design of a building and attributing a symbolic meaning to them is always a potentially dangerous enterprise. In any case, the relationship with calendrics is strongly supported by the archaeoastronomical analysis.

Since the monument has a square base, to define orientation we need only find out the azimuth of the perpendicular to one side (say, the south-east one) looking out (the 90° angles of the base are *not* perfect, in fact, but the error is negligible for our purposes). There is some confusion in the literature about the exact value of this azimuth; I adopt here the value 112° (based on Sprajc and Sanchez Nava 2013, and on a direct measure). Thus, the square base of the Castillo is skewed 22° clockwise with respect to the cardinal directions. Why? By analysing the motion of the sun at Chichen Itza it can easily be verified that the azimuth of the north-west façade (270° + 22° = 292°) fits remarkably well with the setting of the sun on the days of the zenith passages, the 20th of May and the 24th of July. There, is, therefore, little room for doubt that the monument was oriented in such a way that this façade is fully lit by the setting sun on these days. It could be argued, however, that the intended orientation was rather that of the south-east façade, towards the rising sun at azimuth 112° (Milbrath 1988). The horizon is flat, and therefore these dates (18 January and 23 November) correspond to the days opposite the zenith passages, the so called nadirs (on these days, at midnight, the sun is directly overhead in the opposite hemisphere, see Sect. 1.2). However, there is no convincing proof that this

Fig. 9.7 Chichen Itza. The Castillo

"phenomenon" (as a matter of fact, nothing special transpires on the nadir days in the hemisphere where the nadirs do occur) was of interest to the Mayas. The interest in the passage of the sun at the zenith is, on the other hand, very well documented in Mesoamerica; it was studied with appropriate devices (see next section) that enabled astronomers to recognise the transit of the sun directly overhead. Yet another hypothesis suggested is that the intended orientation was that of the diagonals, and not that of the sides. This orientation is sometimes claimed to be along the summer solstice sunrise/winter solstice sunset line. However, the azimuth of the diagonal is $22° + 45° = 67°$, while the azimuth of the sun rising at the summer solstice at Chichen Itza is about 64° 30′, thus with an error of +2° 30′. Also considering that the site is merged with the jungle, with a flat horizon that does not facilitate exact observations, if the alignment was intentional the planners committed a big error, which has to be compared with that—much smaller—of the alignment to the zenith sunset. Therefore, it is difficult to accept that it was the orientation of the diagonal that the architects were interested in at the design stage. Having said that, it is also true that the builders certainly knew that, at the latitude of Chichen Itza', a square building with a side facing zenith sunset would have worked relatively well also as a solstice indicator, with two faces remaining substantially in the shade and two faces fully illuminated at sunrise at summer solstice, and the opposite two at sunset at winter solstice.

Ultimately, it is difficult to escape the idea that the Castillo is a calendrical monument, connected intimately (albeit symbolically) to the regularity of the flow of calendar time and to the repeatability of seasonal, natural time. But the most striking astronomical aspect still awaits. The Castillo was in fact designed as a *three-dimensional* astronomical device, and this feature makes this wonderful monument quite unique.

To understand how this device works, let us observe that, in general, the shadow produced by the north-west corner of the pyramid in the afternoon will be projected onto the north-east face in different ways on different days and at different hours. Generally speaking, the narrow north-east balustrade (the only one bearing serpent heads) could either be in shadow or illuminated. Only as a result of very special positions of the sun (and therefore on special days and hours) will the balustrade be half-illuminated, that is to say, the undulating profile of the corner will project a

Fig. 9.8 Chichen Itza, Spring Equinox. A serpent of *light* and *shadows* descends the staircase of the Castillo ⊙

undulating figure onto it. When this occurs, the light (and shadow) connect with the serpent head at the base, creating an unmistakable image of the descent of a giant rattlesnake from the temple (Fig. 9.8).

The monument has been designed in such a way that this phenomenon takes place—reaching its maximum effectiveness about 1 h before sunset—in the days close to the spring equinox (it was certainly meant to be seen in March, as September is in full rainy season). This calendrical function is also clearly symbolic, since the hierophany cannot be used to establish the equinoctial date with precision; in particular, this also implies that perhaps the days of interest were March 23 and September 21, the quarter days which divide the year into four equal periods of 91 days (Sprajc 2001; Sprajc and Sanchez Nava 2013). Interestingly, focus on the quadripartition of the year is clearly recognisable also in pre-conquest documents, especially in the Dresden codex (Bricker and Bricker 1988) (Fig. 9.9).

The document was purchased by Royal Library at Dresden in 1739, and its provenance is unknown; there is, however, the distinct possibility that it was compiled at Chichen Itza' around 1200 AD, copying material and astronomical data that had already existed for centuries. This idea is fascinating because—at least in my view—a page of the codex (68a, belonging to the so called seasonal tables) seems to contain a reference to the Castillo hierophany. It is a double figure of the Rain God Chac. The two figures sit back to back, on a sort of bench made out of a "sky band", the typical Maya decoration containing symbols of planets and constellations, signalling a celestial context. The two figures are surmounted by what is called a "reversal" glyph, which alludes to the equinoxes; rain is falling on the right-facing Chac only. The reading order of the whole table is from left to right, so that the dry season—symbolised by a dry Chac—precedes the rain season (wet Chac) in the image: to sum up, what is being represented here is the spring equinox as a symbolic divide between the dry and rainy seasons, the transition which was connoted by the serpent hierophany.

It would be quite reasonable at this point to wonder whether the serpent equinox, which descends the Castillo stairway as a symbolic harbinger of the rainy season, is going somewhere. To answer to this question we must broaden our view of the monuments of Chichen Itza' (Carlson 1999). The direction indicated by the stairway where the hierophany occurs crosses another monument of the esplanade, the so-called Venus Platform; it is a sort of square altar, with inscriptions referring to Venus, and serpents' heads. Nearby starts a sacbe, a ceremonial causeway which today can still be run up to the largest of the Chichen Itza' cenotes, the so-called Well of Sacrifice (Fig. 9.10).

The sacred water here is thus the final destination of the god. We know for certain that the well at Chichen Itza had been one of the most important Mesoamerican pilgrimage sites, and underwater archaeology in the cenote has yielded a variety of offerings made over centuries by pilgrims, as well as clear evidence of human sacrifice. However, much is still to be understood about the meaning of this spectacular architectural ensemble. It should be said first of all that it is in fact replicated in the south sector of the city, where there is another radial pyramid, the so called Osario (or Priest's Grave, on account of a cave lying underneath) (Fig. 9.11).

Fig. 9.9 Double figure of the Rain God Chac (redrawn from page 68 of the Dresden codex)

Fig. 9.10 Chichen Itza. The sacred cenote

Fig. 9.11 Chichen Itza. The Osario pyramid

A precise dating of this structure is difficult, but it must be roughly contemporary with, or slightly later than, the Castillo. This pyramid is not oriented like the Castillo though, and no light-and-shadow effect has ever been ascertained there. At any rate, the balustrade "descends" towards another Venus platform, another sacbe', and the second great Chichen Itza' cenote, Cenote Xtoloc. Thus we may see that an architectural replica of the main ceremonial centre of Chichen was created within Chichen itself. But the surprises do not end here. Someone, around 1100 AD, decided to replicate the Castillo in yet another place.

The chosen site was Mayapan, some 100 km to the west, and almost exactly on the same parallel. Spanish sources inform us that the site was founded as a new capital after the fall of Chichen Itza, and that the new capital blossomed for a couple of centuries before being abandoned. These historical accounts are somewhat confused and unreliable, but archaeology has substantially confirmed the period when the city flourished. According to tradition, it was ruled by two elite families, one from Chichen and another from the Uxmal area, and indeed a distinct mixture of styles can be seen in the city centre, which is concentrated around a small cenote. Here, a building in typical Puuc style—replete with impeccable Chac masks on its front—can be seen alongside what looks like a (heavily restored) replica of the Caracol, on a smaller scale (Fig. 9.12). But the masterpiece is the nearby pyramid, which is obviously a near copy —again on a smaller scale—of the Castillo (Figs. 9.12 and 9.13).

Fig. 9.12 Mayapan. A perfectly conserved Chac mask, with the Caracol in the background

Fig. 9.13   Mayapan. The Castillo

The orientation of this building is very interesting. The staircases are orientated to the cardinal points (in the literature the orientation of the sides is given as 5° east, but the monument is not perfectly built as a square and a direct measurement of the staircases I have taken gives a value closer to the cardinal directions). There is no light and shadow effect at the equinoxes, but a diamond-snake effect similar to that of the Castillo occurs also here, for about a month before and after the winter solstice (Aveni et al. 2004). This is, therefore, an example of an architectural replica where the astronomical content was changed; the reason might be a change in the calendar that took place in the same period.

In conclusion, the serpent hierophany at Chichen Itza is another fine example of "testis unus testis nullus" case (excluding Mayapan, since the orientation there is different) where one cannot formally assert the intentionality behind. On the contrary: a rigorous approach—given the impossibility of carrying out a statistical analysis, and the absence of direct written records—would tend to impute the serpent equinox phenomenon to pure coincidence. So it is up to you. You *may* think that the planners of Chichen Itza' devised and constructed the only three-dimensional hierophanic device known from the ancient world, and so, that the light-and-shadow rattlesnake body which takes shape on the balustrade of the temple at sunsets close to the spring equinox and then connects itself with the stone head of a serpent located exactly at the end of the same balustrade, finally moving towards the Venus platform and the sacred way that leads to the most sacred cenote in the Maya world is a willing phenomenon or—well, you may think that all that was just down to pure chance.

9.3 Going Where the Sun Turns Back

There is a strange coincidence connecting the three (chronologically speaking) successive Mesoamerican towns which, as we have seen, are interrelated by historical facts and architectural replicas, viz, Tula, Chichen Itza and Mayapan (Figs. 9.14 and 9.15). The coincidence is that Chichen Itza' and Mayapan lie on the same parallel, and both sit well within 1° of the Tula parallel (latitudes are 20° 40' for Chichen Itza, 20° 38' for Mayapan and 20° 03' for Tula).

This may be due to chance; however there are clear indications that Mesoamerican cultures of central Mexico had somehow come up with the idea of latitude by noticing that the number of days between two zenith passages of the sun is different in different places, and by individuating the place where the zenith passages coincide with the summer solstice, that is, the tropic (actually, the first measurement of earth radius we are aware of, executed by the third century BC scientist Eratosthenes of Cyrene, was based on a similar procedure; recall however that according to all historical evidences, there was no cultural exchange between pre-Columbian people and the classical world). The proof has to be sought in the state of Zacatecas, in North-West Mexico. This area, between the third and the eleventh centuries AD, was inhabited by a people known as Chalchihuites (Kelley 1971). The Chalchihuites were farmers, with a rather crude architecture and relatively simple decorative art; their settlements were located on the hills, in

Fig. 9.14 Tula. Temple of the morning star

Fig. 9.15 Chichen Itza. Temple of the warriors

easy-to-defend positions. Within this cultural scenario, one site should be singled out (Fig. 9.16).

Today called Alta Vista, it is atypically situated on a plain, with no defensive features and bereft of water sources. It exhibits a spectacular series of buildings which pertain to a cultural phase dated—to fix ideas—around 650 AD, during

Fig. 9.16 Altavista. Hall of the columns, with the Cerro Chapin at the horizon

which elaborate ceramics were also produced with sophisticated technique and design. These buildings and the contemporary ceramics are clearly out of place in the local culture. Their design rather recalls central Mesoamerica, including Teotihuacan, and is highly ceremonial in content. We have no traces of an invasion, though; most probably, the two cultures were in contact via trade routes with the American Southwest and the presence of commerce on account of turquoise mines. At a certain point, a more substantial group of central Mexico people, bringing Teotihuacan culture with them, arrived, establishing at Alta Vista a sophisticated ceremonial centre with central Mesoamerican cultural features. The ceremonial centre was built with the use of elaborate construction techniques involving stone and concrete walls. But why did they invest such enormous effort in this very place? The answer, as the reader may have guessed, is probably none other than astronomy (Aveni et al. 1982).

The starting point is the actual position of Altavista, which lies practically on the Tropic of Cancer. To be precise, today the tropic passes some 4.2 km south of the ruins. Due to a slight decrease in the obliquity of the ecliptic, at the time of construction the tropic was in fact some 14 km to the north, but very close to the site anyway, with an error of 7′ in terms of latitude. A second important point is the favourable position of Altavista for making solar observations. At the eastern horizon, the view is occupied by a prominent mountain range, with a distinct central peak, the Picacho. To the south-west, the horizon is taken up by another mountain, El Chapin. Here, near the summit and in view of Altavista, circular petroglyphs chipped into the rocks have been found. Usually called *pecked crosses*, these chipped images immediately suggest a sort of station for the placement of a survey stack or a gnomon. Scores of these have been found in Mesoamerica, in particular at Teotihuacan, where they were probably used for astronomical survey measurements of the town's axes (Aveni et al. 1978). The presence of these etched circles gives a strong hint that the builders of Altavista were performing astronomical observations. This is spectacularly confirmed by the following combination of astronomical alignments. An observer located at the petroglyph on Cerro Chapin can see the June solstice sunrise taking place from behind the Picacho peak, which towers over the town. On the other hand, the very same foresight—the Picacho—was used for equinoctial alignments within the town. A first building showing equinoctial alignment is the so-called Hall of Columns, a square enclosure with 28 columns. This structure was built with painstaking care and accurately oriented with the sides along the inter-cardinal directions; it has virtually no parallel in the area, in either size or orientation (one smaller colonnade hall is to be found at La Quemada, some 150 km to the south-east, but there the orientation is the familiar 17° south of east). The orientation was achieved by means of an elaborate system of survey poles which were later plastered into the walls, and since the building is square, its diagonals are orientated to the cardinal points, so that, in particular, the east-west

Fig. 9.17 The first page of the Codex Fejérváry-Mayer (image in the public domain)

diagonal aligns to sunrise at the equinoxes. There is, therefore, a deliberate link with the quadripartition of the world, confirmed by findings in a multiple burial place, possibly of human sacrifices, located at the north corner. The four corners of the universe were the realm of a Mesoamerican God whom the Aztecs were to call Tezcatlipoca, frequently represented in the codices—for instance in the pre-Hispanic document known as Codex Fejérváry-Mayer (Vail and Hernandez 2010) (Fig. 9.17).

The realm of the God is clearly shown, as the personage is always represented at the centre of quadripartite images (which curiously resemble the plan of the Castillo at Chichen Itza'). Similar images recur in the decoration of Red-on-Brown pottery at Altavista, which show a stylised serpent body motif dividing the interior of shallow bowls and plates into four quadrants. Attention to the cardinal points and equinoxes at Altavista is further corroborated by another structure—called the Labyrinth. The main east-west axis here points directly to the summit of Picacho peak, so that the equinoctial Sun, rising from behind of the peak, suddenly illuminates the halls, creating a spectacular effect.

To sum up, then, Altavista appears to be a cultural-religious outpost of Teotihuacan culture, built at a relevant distance but in line with the high constructive standards of the heartland, in a rural area which was otherwise frequented only for trade and commerce. The site shows a clear intervention by expert astronomers and it is really the only pre-Columbian site where a clear, exclusive interest in the cardinal points and the quadripartition of the universe is exploited in architecture. Its location was carefully chosen to be very close to the Tropic, and—due to its uniqueness—it is almost impossible to think that this was coincidental. Thus we have no alternative but to conclude that the astronomers responsible for Altavista sought out a place where the sun was directly overhead at the summer solstice.

The questions now: why, and how?

The second question is perhaps easier to answer. As already mentioned, the interest in, and observation of, the passage of the sun at the zenith has been well documented in Mesoamerica. Appropriate measuring devices have been found, for example, at Monte Alban, in the Oaxaca valley. Here an underground chamber is connected with the surface by a vertical shaft with a hole in the top. Because of its reduced size, the hole indicates the days of the zenith passages of the sun when the sunbeam falls perpendicularly through the opening. The same kind of measurements can be made by watching carefully the shadows projected by a gnomon around noon, day after day. It is therefore not unreasonable to conjecture that the builders of Altavista combined their basic knowledge of the fact that in going north, the interval of days between the zenith passages diminishes, with perhaps impromptu observations reported by travellers about the existence of places to the north where the phenomenon no longer occurs. In any case, individuating the Tropic so precisely could not have been an easy matter. Why did they persist in such an endeavour?

It is difficult to assess how the existence of the line of the Tropic might have been perceived. It is natural to think that the astronomers were aware that the changing behaviour of the sun when going north is due to the fact that the earth's surface is not flat but spherical. At least in my view, no serious sky-watcher can really believe the Earth to be a flat surface: for any astronomer the Earth is at the very least a disk, since one only has to watch a lunar eclipse to see the curved rim of the earth progressively covering our satellite; furthermore, tall objects on the horizon (like ships' masts) first become visible in their higher parts, and if this might pass unnoticed by most people, it most certainly was noticed by people who spent their life gazing at the horizon; finally, for observers in tropical areas it is extremely hard to explain the variation in the number of days between two zenith passages in relation to the place if all places are located on the same flat surface, and it is equally difficult for anyone to explain the differences in altitude of the north celestial pole in different locations.

Of course however, this does not mean that the Earth's spherical nature was common knowledge, and cosmologies usually employed neater, simpler concepts that could be understood at a religious level. The Tropic might have been understood as a boundary beyond which the sun could no longer reach up to the zenith at noon. The construction of Altavista in this case would have had a foundational aspect: fixing the correct location of the zenith. Interestingly enough, in some depictions of the God responsible for the quadripartition of the universe, the cardinal points are represented, together with other paraphernalia, as rising trees. Along with an image for each of the four of them, a fifth direction is also shown and represented as a tree arising from "the centre". So the overhead direction might have been identified as arising from a "centre" and this centre might have been indicated by Altavista (Aveni 2001).

Whatever significance this intriguing place might have had for its builders, Altavista is a striking example of a sacred landscape whose localisation was literally dictated by astronomical considerations. Incredible as it may seem, there is the distinct possibility that it was not the unique site which was frequented as a place of worship by a culture expert in astronomy for its proximity to the tropic. Another may in fact be Necker Island, a small island in the North-western Hawaii (Ruggles 2005). Necker is little more than a rock emerging from the sea almost exactly on the tropic at 23°34′35″N. It is difficult to believe that the Hawaiians could live here in the long-term, but the place was frequented several times as more than 30 stone shrines have been found there. So the place was a ceremonial site perhaps frequented for its proximity to the tropic (accurate knowledge of astronomy by the Hawaiians for navigational purposes is beyond question).

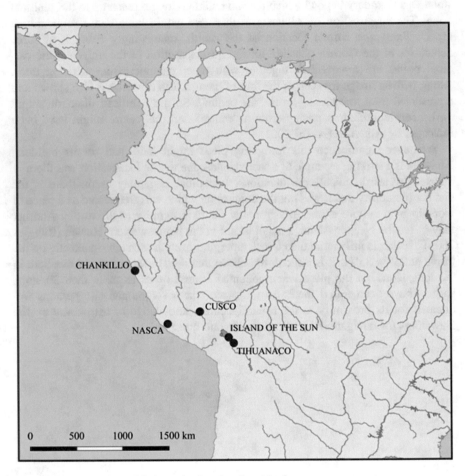

Map of South America, with the main sites mentioned in the text

9.4 Pillars of the Sun

In this section we shall discuss some aspects of one of the most complex cultures ever, that of the Incas. In dealing with such a complex subject, it is important to stress that the Incas—whose civilization was destroyed by the Conquistadores in the first half of the sixteenth century—were only the final phase of a series of cultures which started building monumental architecture in Peru as far back as the third millennium BC (Moseley 2001). The Incas were therefore the heirs to a millenarian legacy of knowledge and craftsmanship. Their ascent was irresistible: in the thirteenth century they were just one among several groups of people living in the Cusco area (we do not know what name they gave themselves, as Inca was only the way they designated their ruler). In the following 200 years they succeeded in building up a substantial empire stretching across the territories which today span from Colombia to Argentina.

The Inca state was called *Tahuantinsuyu* or "The Four Parts of the Earth". It was organised according to a centralised system, and ruled with the use of an official language, *Quechua*. Governance of the country was in the hand of a rigid hierarchy, with the noble Inca families, residing in the capital, Cusco, on top. In Cusco, the state archives recorded meticulously a huge amount of data (such as the taxes paid in every part of the empire) using devices, called *Quipus*, which consisted of a cluster of ropes tightened to a main one and laid out like a tree (Fig. 9.18). The ropes could be of different colours, length and dimensions, and were tied with a series of different knots, so that masses of different information could be stored (Urton 2003). It seems likely that these devices were also used in some way as a form of written language, to record, for instance, short poems, but the small number of surviving Quipus and the paucity of information passed down in their regard does not permit, at present, a full understanding of their use.

The Incas were extraordinary builders: many of their constructions employed the so-called polygonal technique, assembling huge blocks—shaped like irregular polygons—one on top of the other using apparently crazy, multi-angled, but perfect joints. This kind of monumental architecture had a fundamental symbolic value: official constructions in stone—in particular ruler's palaces and estates—repre- sented the "materialisation" of history, a way of asserting eternal rights for the ruler's descendants and family (Niles 2004). Another masterpiece of Inca engi- neering is the network of roads which criss-crossed the entire empire. These roads were traversed rigorously on foot, since the Andean beasts of burden, the llamas, are not suited to pulling carts and categorically refuse riders (the complaint that the Incas did not invent the wheel, which can be read in many sources, being utter nonsense—they just had no need for it).

Religion was a seminal unifying element in the Inca state, as well as an effective instrument in the management of centralised power. In their cosmological vision, the Incas traced their origin to Lake Titicaca. According to the legend, the God-Creator Viracocha created the world: the celestial bodies, the earth and the sun, which rose for the first time from a rock in the Titicaca Island thereby called Island

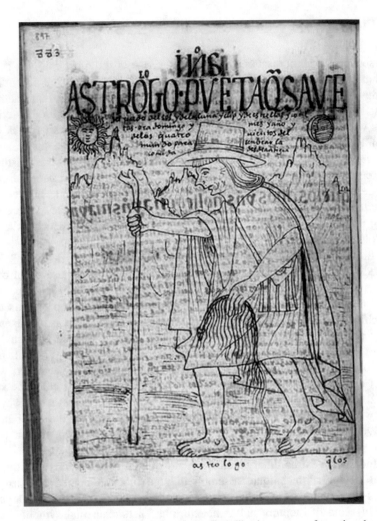

Fig. 9.18 Inca astronomer carrying Quipu and a fork-like instrument, from the chronicle by G. Poma de Ayala (Image in the public domain)

of the Sun. The first Inca, Manco Capac, was considered the "son of the sun", so that his right to rule came directly from the gods. The close relationship between rulers, gods and nature was reflected in the primary aspects of religious life. Deceased kings were mummified and worshipped, together with a series of objects, places and natural phenomena whose cult was rigorously organised and managed: in a sense the entire landscape was considered as sacred. Beyond the level of the sacred landscape there was a superior level—the celestial one—and an inferior one —that pertained to the earth; these three levels were connected by the flowing of the rivers from Earth into the great celestial river, the Milky Way or *Mayu*. The Inca world is thus a paradigmatic example of a tiered, ordered cosmos. The three levels

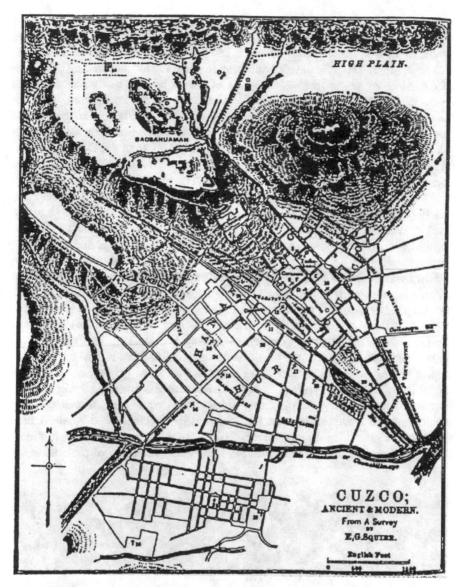

Fig. 9.19 Plan of Cusco drawn by E. Squier in 1860. The profile of a four-legged animal aligned northwest-southeast is easily recognizable (image in the public domain)

of the Cosmos and the four parts of their state converged into the very heart of the empire, the navel of the Inca world: Cusco (Fig. 9.19).

According to the descriptions left by Spanish chroniclers, Cusco was magnificent. Since the post-conquest buildings were superimposed on the pre-existing, megalithic Inca buildings, the urban layout of the original city is still perfectly visible. It is quite a strange layout: it is not orthogonal, nor circular, it has no real geometrical design, yet it is far from being haphazard. The city—the only one of its kind in the world—was planned in a manner that aimed to make it look, when seen from above, like the profile of a Puma (Rowe 1967; Gasparini and Margolies 1980). This curious effect was achieved by fixing the town's boundaries in a specific location, taking advantage of natural geographical features. The back of the feline is in fact outlined by two rivers, the Tullumayu and the Vilcanota, which converge to form the tail, while the head is represented by a hill, the Sacsahuaman. The hill boasts an unequalled masterpiece of Inca megalithic architecture, a parallel series of three zig-zag walls whose blocks weigh anything up to 300 tons, which—despite the Conquistadores' childish interpretation of it as a fortress—was clearly built for symbolic reasons, and perhaps represented another sacred animal, the serpent (Fig. 9.20).

All in all, Cusco is a unique case (to my knowledge) of an urban plan devised to satisfy purely symbolic criteria. But this was not enough, since for the Incas the level of celestial things and of the sacred landscape mirroring them were undistinguishable, and therefore not only architecture, but also religion and astronomy, were interrelated in a way that is hard for us to understand because it is so far from our own way of thinking and behaving. The idea of creating a town in the shape of a Puma is an example: the puma had been—together with the eagle—the symbol, the insignia, of the ruler, and Cusco had been the capital, the ruler's elected residence since the mythical times of the first Inca. As such, it was also the true heart of the Inca world, the place from which everything had to start. This concept was developed in the division of the town and its surroundings into radial sectors called *ceques*. Along each ceque there were many sacred sites, or *huacas*, of various kinds, including stones, springs, caves and other places considered sacred for a number of reasons. The people living along each Ceque were in charge of the maintenance of these huacas, and this maintenance included both administrative tasks, such as checking on the correct distribution of the waters, and ritual procedures, like bringing offerings on given days. We know about the whole structure of

Fig. 9.20 Cusco. The central sector of the Sacsahuaman "fortress". Several blocks weigh more than 300 tons each

the Ceque system thanks especially to the detailed report written by the chronicler Bernabè Cobo, who says that 41 ceque lines existed, and that 328 huacas were located along them (the ceques were not perfectly straight lines, as their direction could vary a little from one huaca to another); many huacas were later identified by archaeological research (Rowe 1979; Aveni 1981, 1996; Zuidema 1964, 1988; Bauer 1998).

Looking at the radial ordering of space within the ceque system one is struck by the idea that it is undeniably similar to the mental organisation lying behind the Quipus. In fact, a map of the Cusco ceques resembles a Quipu spread on the ground, a sort of *gigantic replica* of a Quipu (it might not even be a symbolic replica of a general Quipu but rather the gigantic copy of a specific one which still awaits to be read, see Magli 2005). In any case, it is likely that the recording system based on *Quipus* was also used to keep track of scientific observations of the sky, just as happened with the Maya codices, and indeed, in a chronicle written shortly after the conquest by Guaman Poma de Ayala, an Inca astronomer is depicted with a fork-like observation instrument in one hand and a Quipu in the other hand. Clearly, then, it would be natural to wonder if the ceques were astronomically oriented. The broad answer to this question is negative, but very important huacas in the system certainly corresponded to astronomical observations. These huacas are described by Cobo and other writers as being stone pillars located at the horizon and dedicated to the sun. At least four sets of such pillars must have existed on the Cusco horizon; for instance a huaca called *Quiangalla* is described as "a hill on which there were two markers or pillars which they regarded as indication that, when the sun reached there, it was the beginning of summer".

The problem of identifying the pillars is a fascinating one for archaeoastronomy, since they would be one of the few known purely astronomical devices of ancient antiquity, what we could really call "an ancient observatory" in the strictest sense. Unfortunately, and in spite of extensive research, the stone pillars of Cusco, or at least their foundations, have never been identified, and the chances of doing so are becoming increasingly slim owing to increasing urbanisation and the looting of Inca sites. The reason for the failure is that all the Inca haucas were systematically sought out and destroyed—as symbols of the old religion—in the two centuries following the conquest. Clearly—and unfortunately—the sun pillars must have been among the most visible of the huacas. Also uncertain is the exact point of observation from the town, although it must have been at the main temple or at the plaza nearby. As a consequence of this lack of evidence, there is some debate about the dates which were defined by the pillars, although the two solstices—corresponding to two of the most important rituals of the year, Capac Raymi (winter solstice) and Inti Raymi (December, and thus summer, solstice) and a date in August seem certain. August is the month of one of the two nadir passages at the latitude of Cusco, so that it has been proposed that the Incas were interested in this passage; there is, however, no historical evidence for this interest and, after all, August was very important in the agricultural cycle (Zuidema 1988; Bauer and Dearborn 1995).

All in all, the situation appears quite frustrating. Yet, in recent years, unexpected discoveries have been made outside Cusco that have dramatically changed this.

First of all, interesting discoveries have been made on the Island of the Sun (Dearborn et al. 1998; Bauer and Stanish 2001). Being the place of mythical origin of the Incas, in this tiny island of the Titicaca Lake there was a complex and very important sanctuary, managed by the state and visited by pilgrims from every part of the empire. The sanctuary was perceived as the most sacred destination "to the south-east" (in relation to Cusco), in a similar way perhaps as was Machu Picchu, the enigmatic town perching like an eagle's nest on a mountain at a turn of the Urubamba valley, to the north-west (Magli 2010) (Fig. 9.21). The pilgrims gathered on the shores of a town today called Copacabana, were ferried to the island and then followed a path that led them to see the "footprints of the sun" we already mentioned in Sect. 5.5 and finally the sun rock itself, the place where the sun was purported to have been born. The esplanade directly in front of the sacred rock was probably reserved for the elite, while ordinary people gathered on the low hills just behind. From all these locations it was possible to see the sun at the winter (June) solstice rise from behind the sacred rock—thus symbolising a re-enactment of the first dawn—and set on a hill to the north-west, the Tikani ridge. Here, the stone foundations of two structures have been recovered, and a simulation carried out with two modern posts placed in these foundations has shown that they were probably the positions of two horizon pillars that acted as markers of the June solstice sunset as viewed from the sacred rock area (interest in the June solstice is also apparent at Macchu Picchu, see Dearborn and White 1983; Ziegler and Malville 2003).

Fig. 9.21 Machu Picchu. General view ▶

An even more striking discovery relating to the winter solstice was made in the Urubamba valley. Along the course of the river, scattered here and there going from Cusco up to the ridge surmounted by Machu Picchu and her recently re-discovered sister Llactapata, several Inca complexes can be seen (Protzen 1993; Burger and Salazar-Burger 2004; Malville et al. 2004; Reinhard 2007). Many were private properties of the rulers, types of royal estates, used—or at least so it is thought— both for private purposes and official ceremonies. One such complex was called Quespiwanka and is located near a village called, like the river, Urubamba. Quespiwanka was a royal property of Huayna Capac, the last Inca emperor before the arrival of the Spaniards, which occurred during the reign of his son, Atahualpa. It is a relatively small property furnished however with a vast plaza, constructed in such a way as to leave a huge white boulder of granite (after which the palace was probably named) fairly conspicuous. Standing in the plaza and looking at the

Fig. 9.22 Urubamba. The Inca pillars on the mountain ridge. *Courtesy* Mike Zawalski ⊙

eastern horizon, a imposing mountain ridge can be seen, towering some 1000 m above the village below. Looking carefully, the existence of two artificial protuberances on the ridge can be discerned (Fig. 9.22).

These two protuberances are actually two Inca towers, whose existence has always been known about. They stand at a distance of 2.2 km from the courtyard as the crow flies, stand 35.3 m apart and are carefully constructed on a common artificial platform; one of them is almost intact, and it has a height of 4.3 m and a base measuring 1.5 × 3.3 m. It is, of course, unthinkable that these buildings were unknown to the Spaniards at the time of the destruction of the Inca huacas, but Quespiwanka was distant enough to escape the attention of the inquisitors in Cusco, and the palace rapidly fell into ruin after the conquest and providentially avoided destruction (later, one of the towers was fitted with a Christian cross on its top). Incredible as it may seem, however, up to some years ago nobody had ventured to probe the meaning of these curious features. A meaning which, I think, you might just guess.

Standing in the courtyard—at the white boulder, say—the azimuth of the mid-point of the pillars is 56° 14′; their mean altitude is 23° and the result is that they act as precise markers of the winter solstice sunrise (Zawaski and Malville 2010; Malville et al. 2009). The phenomenon is visible from the whole palace area, thus also from beyond the courtyard walls, access to the latter probably being reserved only to the elite. The fact that the pillars are built on a terraced platform is significant, because this distinguishes them as not merely being a signpost device, but a huaca to all effects.

A further, striking discovery has been made in recent years in another Peruvian site, Chankillo, a ceremonial centre that was active between 350 and 150 BC circa (Ghezzi and Ruggles 2007). The main constructions at Chankillo are a massive fortified temple and a large ceremonial area with several buildings. A natural hill in the shape of a rib traverses the site; on the hill, a row of thirteen rectangular towers was constructed. The hill forms a natural horizon which appears with a serrated profile, due to the presence of the towers which run in an approximate north–south line. They are regularly spaced, with gaps about five meters wide. Each tower has a staircase on both the sides along the hill, so it was possible to walk along the ridge of the hill, up and down each tower, perhaps on the occasion of processions (Malville 2011) (Fig. 9.23).

An archaeoastronomical analysis of this enigmatic place shows that the towers were used as a device for observing the cycle of the sun, which was seen to rise behind them over the course of the year. The observation point is identifiable in a building located to the west of the towers, between them and the fortified temple. This building possesses a carefully constructed long corridor which leads to a (doorless) opening directly facing the thirteen towers. The view towards the hill actually takes in not 12 but 13 indentations, because of a hill located 3 km away, which rises just to the left of the northernmost tower. The hill thus forms a further obstacle behind which the sun can be seen to rise. As viewed from the opening, the declinations associated with the first and the last of the gaps correspond very well to the maximal and minimal declinations of the sun; progressively over six months, the sun thus rises behind successive towers (however, no further special dates, like equinoxes or zenith passages, seem to be highlighted) (Fig. 9.24).

Fig. 9.23 Chankillo. Panoramic view of the thirteen towers. *Courtesy* Ivan Ghezzi ▶

Fig. 9.24 Chankillo. The June (winter) solstice sunrise occurs between Cerro Mucho Malo and Tower 1, as viewed from the western observation point. *Courtesy* Ivan Ghezzi

References

Ashmore, W. (1991). Site planning principles and concepts of directionality among the ancient Maya. *Latin American Antiquity, 2,* 199–226.

Aveni, A. F. (1981). Horizon astronomy in Incaic Cusco. In R. Williamson (Ed.), *Archaeoastronomy in the Americas* (pp. 305–318). CA: Ballena, Los Altos.

Aveni, A. F. (1996). Astronomy and the ceque system. *Journal of the Steward Anthropological Society, 24*(1/2), 157–172.

Aveni, A. F. (2001). *Skywatchers: A revised and updated version of skywatchers of Ancient Mexico.* Austin: University of Texas Press.

Aveni, A. F., & Hartung, H. (1978). Three Maya astronomical observatories in the Yucatan peninsula. *Interciencia, 3,* 136–143.

Aveni, A. F., & Hartung, H. (1986). Maya city planning and the calendar. *Transactions of the American Philosophical Society, 76,* 1–79.

Aveni, A. F., Gibbs, S. L., & Hartung, H. (1975). The Caracol tower at Chichen Itza: An ancient astronomical observatory? *Science, 188,* 977–985.

Aveni, A. F., Hartung, H., & Buckingham, B. (1978). The pecked cross symbol in ancient Mesoamerica. *Science, 202,* 267–279.

Aveni, A. F., Hartung, H., & Kelley, J. C. (1982). AltaVista (Chalchihuites), astronomical implications of a Mesoamerican ceremonial outpost at the Tropic of Cancer. *American Antiquity, 47,* 316–335.

Aveni, A. F., Milbrath, S., & Peraza Lope, C. (2004). Chichen Itza's legacy in the astronomically oriented architecture of Mayapan. *Anthropology and Aesthetics, 45,* 123–143.

Bauer, B. (1998). *The sacred landscape of the Inca: The cusco ceque system.* Austin: University of Texas Press.

Bauer, B., & Dearborn, D. (1995). *Astronomy and empire in the Ancient Andes.* Austin: University of Texas Press.

Bauer, B., & Stanish, C. (2001). *Ritual and Pilgrimage in the Ancient Andes: The Islands of the Sun and the Moon.* Austin: University of Texas Press.

Bricker, V. R., & Bricker, H. M. (1988). The seasonal table in the dresden codex and related almanacs. *Journal for the History of Astronomy Archaeoastronomy Supplement, 19,* S1.

Burger, R., & Salazar-Burger, L. C. (2004). *Machu Picchu: Unveiling the Mystery of the Incas.* New Haven and London: Yale University Press.

Carlson, J. B. (1999). Pilgrimage and the equinox "serpent of light and shadow" phenomenon at the Castillo, Chichen Itza. *Yucatan. Archaeoastronomy, 14*(1), 139–152.

Coe, M. (2001). *The maya.* NewYork: Thames and Hudson.

Dearborn, D., & White, R. (1983). The Torreon of Machu Picchu as an observatory. *Archaeoastronomy, 5,* S37–S49.

Dearborn, D., Seddon, M., & Bauer, B. (1998). The sanctuary of Titicaca: Where the sun returns to earth. *Latin American Antiquity, 9,* 240–258.

Gasparini, G., & Margolies, L. (1980). *Inca architecture.* Bloomington: Indiana University Press.

Ghezzi, I., & Ruggles, C. (2007). Chankillo: A 2300-year-old solar observatory in Coastal Peru. *Science, 315*(5816), 1239–1243.

Kelley, J. C. (1971). Archaeology of the northern frontier: Zacatecas and Durango. In R. Wauchope (Ed.), *Handbook of Middle American Indians* (vol. 11, pp. 768–801). Austin: University of Texas Press.

Kowalski, J. K. (2012). Twin tollans: Chichén Itzá, Tula, and the Epiclassic to early Postclassic Mesoamerican World. Harvard University Press.

Magli, G. (2005). Mathematics, astronomy and sacred landscape in the Inka Heartland. *Nexus Network Journal—Architecture and Mathematics, 7,* 22–32.

Magli, G. (2010). At the other end of the sun's path. A new interpretation of Machu Picchu. *Nexus Network Journal—Architecture and Mathematics, 12,* 321–341.

Malville, J. M. (2011). Astronomy and ceremony at Chankillo: an Andean perspective. In C. L. N. Ruggles (Ed.), *Archaeoastronomy and ethnoastronomy: building bridges between cultures* (pp. 154–161). Cambridge: Cambridge University Press.

Malville, J. M., Thomson, H., & Ziegler, G. (2004). El observatorio de Machu Picchu: Redescubrimiento de Llactapata y su templo solar. *Revista Andina, 39,* 9–40.

Malville, J. M., Zawaski, M., & Gullberg, S. (2009). Cosmological motifs of Peruvian huacas. In J. Vaiskunas (Ed.), *Astronomy and cosmology in folk traditions and cultural heritage* (vol 10, pp. 175–182). *Archaeologia Baltica.* Klaipeda: Klaipeda University.

Milbrath, S. (1988). Astronomical images and orientations in the architecture of Chichén Itza. In A. F. Aveni (Ed.), *New Directions in American Archaeoastronomy* (pp. 57–79). Oxford: bar (British Archaeological Reports, International Series, 454).

Milbrath, S. (1999). *Star gods of the Maya: Astronomy in art, folklore, and calendars.* Austin: University of Texas Press.

Moseley, M. (2001). *The incas and their ancestors.* London: Thames and Hudson.

Niles, S. A. (2004). The shape of inca history. Narrative and architecture in an Andean empire. University of Iowa Press.

Protzen, J. P. (1993) *Inca architecture and construction at Ollantaytambo.* Oxford University Press.

Reinhard, J. (2007). *Machu picchu: exploring an ancient sacred center.* New York: Cotsen Institute of Archaeology, NY.

Rowe, J. (1967). What kind of a settlement was Inca Cuzco? *Ñawpa Pacha, 5*(1967), 59–70.

Rowe, J. (1979). Archaeoastronomy in Mesoamerica and Peru. *Latin American Research Review, 14,* 227–233.

Ruggles, C. L. N. (2005). *Ancient astronomy: An encyclopedia of cosmologies and myth.* London: ABC-CLIO.

Schaefer, B. (2006). Case studies of the three most famous claimed archaeoastronomical alignments in North America. In T. Bostwick, & B. Bates (Eds.), *Viewing the Sky Through Past and Present Cultures: Selected Papers from the Oxford VII International Conference on Archaeoastronomy* (pp. 27–56). Phoenix: City of Phoenix Parks and Recreation Department.

Schele, L., Freidel, D., & Parker, J (1995). *Maya cosmos.* Quill. New York.

Sprajc, I. (1993). Venus orientations in ancient Mesoamerican architecture. In C. L. N. Ruggles (Ed.), *Archaeoastronomy in the 1990s.* Loughborough: Group D Publications.

Sprajc, I. (2001). *Orientaciones astronómicas en la arquitectura prehispánica del centro de México.* México: Instituto Nacional de Antropología e Historia (Colección Científica 427).

Sprajc, I., & Sanchez Nava, P. F. (2012). Orientaciones astronomicas en la arquitectura maya de las tierras bajas: nuevos datos e interpretaciones. In B. Arroyo, L. Paiz, & H. Mejia (Eds.), *XXV Simposio de Investigaciones Arqueologicas en Guatemala* (Vol. 2, pp. 977–996). Tikal: Instituto de Antropologia e Historia.

Sprajc, I., & Sanchez Nava, P. F. (2013). Astronomia en la arquitectura de Chichen Itza´: una reevaluacion. *Estudios de Cultura Maya, 41,* 31–60.

Urton, G. (2003). *Signs of the Inka Khipu: Binary coding in the andean knotted-string records.* Austin: University of Texas Press.

Vail, G., & Hernandez, C. (2010). *Astronomers, Scribes, and priests intellectual interchange between the Northern Maya Lowlands and Highland Mexico in the Late Postclassic Period.* Dumbarton Oaks: Dumbarton Oaks Res. Publ.

Zawaski, M., & Malville, J. M. (2010). An archaeoastronomical survey of major Inca sites in Peru. *Archaeoastronomy: Journal of Astronomy in Culture, 10,* 20–38.

Ziegler, J., & Malville, K. (2003). Machu Picchu, Inca Pachacuti's Sacred City: A multiple ritual, ceremonial and administrative center. *Inkaterra, 1,* 1–5.

Zuidema, R. T. (1964). *The ceque system of cusco: The social organization of the capital of the Inca.* Leiden: Brill.

Zuidema, R. T. (1988). The pillars of Cusco: Which two dates of sunset did they define? In A. Aveni (Ed.), *New directions in American archaeoastronomy* (pp. 143–169). BAR International Series 454, Oxford.

Map of the Mediterranean, with the main sites mentioned in the text

Chapter 10
The Classical World

10.1 Houses of the Gods

The classical world, or classical antiquity, is a traditional all-embracing definition describing the cultures of the Mediterranean area from the Ancient Greek civilisation (eighth century BC) to the end of the Roman Empire. It thus includes Classical Greece, Hellenism, and Rome. This chapter is devoted to this broad historical period, and will present some examples of applications of archaeoastronomical ideas and techniques, beginning with the Greek temples of Sicily.

Greek sacred architecture is fêted for its hundreds of magnificent temples, built over the course of several centuries, from the seven century BC onward (Lawrence 1996). Leaving aside regional and chronological distinctions in the layout and in the column orders, which make the precise classification of Greek temple architecture somewhat problematic, these buildings were always based on the same conception: a imposing rectangular construction adorned with columns on the façade. Although in many cases the presence of columned porticoes on all sides made the view of the structure enjoyable from all directions, the main principle always remained the same: a Greek temple was meant to occupy a natural place with an obviously man-made feature, and it was to be admired from the outside only.

Admission was reserved to priests and to the privileged few, and public rites were celebrated outside, in front of the temple, which in many cases was equipped with an altar and a platea (religious occasions included festivals, processions and long rituals). The interior of the temple was, strictly speaking, the home of the god. The god's domestic welfare (hence, the beauty and decorum of the building, correct insertion in the landscape, regular giving of daily offerings) was fundamental to assure benevolence and protection to the community. The cult image, located in the central place of the temple, was in many cases an out-and-out masterpiece, like the famous ivory-and-gold statues of Zeus at Olympia and of Athena in the Parthenon in Athens.

G. Magli, *Archaeoastronomy*, Undergraduate Lecture Notes in Physics,
https://doi.org/10.1007/978-3-030-45147-9_10

The positioning of Greek temples has been the subject of some interesting scholarly research. For instance, a connection between the terrain on which the temple is erected and a related deity has been suggested (Retallack 2008). In 84 temples where the divinity worshipped is known, we find an association between terrain and specific gods (the analysis of temple orientations attempted in the same work, however, is extremely naïve and unacceptable). Some of these associations appear quite natural: underworld divinities are associated with rock, gods of fertility and agriculture like Demeter and Dionysos are associated with cultivated land, and so on; other are less easy to grasp.

Then there is the landscape. In his huge erudite tome, Scully (1962) was the first to stress the importance of *Genius Loci*. His work pioneered research on the Archaeology of the Landscape, pursuing the idea that landscape and temples formed an architectural unit that was characterised in accordance with the specific god worshipped (Fig. 10.1). Scully traced back the origin of these ideas to the pre-Greek, Minoan civilization, and focused his attention on the fact that many sacred places were built within sight of distinctive clefts, distinguishable as double peaked arms, horns, and/or breasts. The prototype was considered the palace of Knossos on Crete, where the court opens to the clefts of Mount luktas, traditionally identified as the profile of a human head. Scully found several other examples, both on Crete and on mainland Greece, for instance, at the Eleusis sanctuary and at the Acropolis at Athens, which has a view of the peaks of Mount Hymettos.

Fig. 10.1 Segesta. The temple in its landscape

Scully's arguments have been criticised however, mostly on the grounds that significant topographical features of this kind are relatively common in Greece. These criticisms—besides being rather vague—become invalid when it is possible to show that the topographical features in question had cultural significance (Hannah 2013). In any case, neither the choice of the terrain nor that of the landscape generally implies a specific orientation, so that the matter of the orientation of the Greek temples deserves to be dealt with on its own.

The orientation of a Greek temple is defined as the direction of the main axis from inside looking out, which is the direction in which the statue of the god was in principle looking, as well as being the direction along which the sun would illuminate the façade, which, as we have seen, was the scene for rites and celebrations taking place outside the temple. The majority of these monuments face east, mostly within the arc of the rising sun (Nissen 1873; Penrose 1897; Dinsmoor 1939). However, recent research has shown that eastern orientation is not the universal key to Greek temples, as was previously believed (Boutsikas 2009). The scenario is actually much more complex, and further analyses are required to take local considerations and traditions into account (Liritzis and Vassiliou 2003; Salt and Boutsikas 2005; Boutsikas and Ruggles 2011; Liritzis and Castro 2013). Here, we shall concentrate on a specific case, that of the Greek temples of Sicily.

The foundation of the Greek colonies in the Mediterranean began as far back as the eighth century BC. They were set up for commercial reasons, but also, in some cases, for political or demographic reasons. Southern Italy was one of the preferred destinations, and Greek culture was exported into Sicily with the foundation of important colonies such as Syracuse, Akragas (Agrigento), and Gela. Hand in hand with culture, the Greeks exported their religion, and consequently several magnificent temples were constructed in the years up to 242 BC, when Sicily became a Roman province. Even today, many Greek temples of Sicily stand as unequalled masterpieces; among them Segesta, the Temples of Selinunte, and the monuments of the Valley of the Temples at Agrigento (Figs. 10.2 and 10.3).

The orientation of the temples of Sicily demonstrates a very clear pattern (Aveni and Romano 2000; Salt 2009). It has been determined that 38 out of 41 measured temples are oriented within the arc of the rising sun. This sample is virtually exhaustive for all the existing monuments, and clearly we have no need of any statistical analysis to conclude that orientation to the rising sun was intentional (the enthusiastic reader may check that the level of confidence is greater than 8σ). We can, therefore, safely assert that the builders intentionally oriented these monuments towards the point on the horizon where the sun rises on a specific day of the year (of course, if the temple is not aligned to a solstice, there are 2 days during which the same phenomenon occurs). This is an important conclusion which we can add to our sum of knowledge of these wonderful monuments (Fig. 10.4).

However, in a way, we are only at the beginning. As it happens, there is no specific concentration of data, for instance, around the solstices or the equinoxes, or other dates for that matter—so how was the alignment chosen? Was it the day of foundation of the temple, or the day of the feast of the god, or what? Perhaps there was a tradition passed down from the original town of provenance? So far, attempts

Fig. 10.2 Selinunte. Temple E, interior ⊙

Fig. 10.3 Akragas (Agrigento). Concordia temple

Fig. 10.4 Akragas (Agrigento), spring equinox. The sun rises in alignment with the axis of the Concordia temple. (*Courtesy* Andrea Orlando)

to gain more insight into this problem—for instance, by investigating on possible groupings for patron deities—have not been successful. Matters are complicated by the fact that the calendars in use in Greek towns were luni-solar, so that alignments based on feast days would not have been calendrically effective in relation to the timing of the rituals carried out (presumably at dawn) in front of the temple. The orientation also appears somewhat unusual when one looks at comparable families of monuments, for instance the Italic temples (temples of the peoples inhabiting continental Italy before the Roman conquest, like the Samnites) which are orientated to the sun ascending in the sky, and the Etruscan temples, which are mostly oriented to the sun ascending or descending in the sky, that is, between the winter solstice sunrise and the winter solstice sunset (Aveni and Romano 2000). It should also be noted that a stellar orientation cannot be distinguished from a solar one if it occurs within the solar arc. Thus all Greek temples oriented to the rising sun also happen to be broadly oriented towards the constellations in which the sun was rising, and can on occasion be accurately oriented to specific stars of such constellations as well as to other stars that had the same declination at the time of construction. A possible, specific interest by the builders in this kind of stellar target must be investigated separately case by case (see e.g. Boutsikas and Ruggles 2011).

So yes, the Greek temples of Sicily are an interesting example of a problem for which we do have authoritative mathematical information regarding orientation, but we have no clear idea as to how this information should be interpreted. Furthermore, we are already acquainted in this book with the fact that the relationship between architecture, astronomy and landscape can be expressed in more complex ways than mere astronomical alignments, and this appears to occur in Sicily as well (Hannah et al. 2017, 2018). For example, among the most spectacular Greek temples of Sicily stand the three built on the rock ridge overlooking the town of Selinunte from the east, labelled E, F and G. The three temples are virtually parallel to each other and share similar orientations to the rising sun at the end of February/mid October with an azimuth of ~96°. However, a closer inspection of temple E reveals that the axis points to a particular feature on the distant horizon: a prominent peak called Mount Nadore. This hill was inhabited since the Iron Age and acts as a topographical marker at the horizon for the nearby Mount Kronio, which towers over the seaside town of Sciacca. Kronio was sacred since Neolithic times thanks to the presence of hot springs, and was perhaps devoted to Herakles. So the peak is a likely candidate as the topographical target which influenced the positioning and the orientation of the temple (Hannah 2013). Further, the longitudinal roads (Stenopoi) of the Selinunte grid plan all point to the very same target.

10.2   The City of the Lion

Alexander III, son of Philip II, King of Macedon, was born in 356 BC in Aigai (today Vergina), the Macedonian capital. Tutored by Aristotle, the young prince soon displayed extraordinary strategic capacities. Aged only 20, he succeeded his father to the throne, and in 10 years, with a series of spectacular campaigns, he defeated the Persians and created an enormous empire, extending as far as Egypt to the south and western India to the east (Lane Fox 2004).

The life of this extraordinary man was not destined to last long: Alexander died in Babylon, in 323 BC. He had no obvious direct heir, since his wife Roxanne was still pregnant with their first son Alexander IV at the time of his death. After birth, the baby and his father's half-brother Philip were appointed joint kings. However, unrest rapidly set in and gave rise to state assassinations and to a lengthy war, at the end of which the Hellenistic world emerged split into four kingdoms ruled by four regnal families: Ptolemaic Egypt, Seleucid Asia, Attalid Anatolia, and Antigonid Macedon.

Alexander was a man of exceptional character and learning. Bearing witness to this is the enormous (over twenty) number of cities founded during his campaigns. The foundation of towns was of course a part of conquest policy, and indeed some cities were simply permanent military camps set up to accommodate part of the army so that they could control the territory. However, many others were founded with more noble designs in mind. The crowning glory of Alexander's new cities is without doubt Alexandria, Egypt, founded in 331 BC (Bagnall 1979). The foundation of Alexandria was a symbolic act, aimed at celebrating Alexander's power and divine nature. As a consequence, ideology is reflected in the city design, which was related to the idea of the ideal town as theorised by Socrates and Plato (Castagnoli 1971). The rigorous order of the orthogonal plan was meant as a reflection of the cosmic order imposed by Alexander as successor to the Pharaohs. In this respect it should be noted that just before the foundation of Alexandria the king had probably consulted the most important oracle in Egypt, the Ammon oracle at Siwa, in an obvious parallel with the Greek tradition by which the founder of a new colony would first consult the oracle at Delphi. According to the same tradition, after his death, the founder would assume the role of divinised ancestor for the town, and would be buried in a suitably devised monumental sepulchre; regrettably, Alexander's tomb—whose existence in Alexandria has been recorded in many historical sources—has never been identified.

The town was conceived on the basis of a main longitudinal axis, later called Canopic Road; the most important transverse axis was the Eptastadion, connecting the mainland with the isle of Pharos (Fig. 10.5). The Canopic Road acted as a sort of elongated centre, with the main buildings distributed along it. Such a longitudinal axis is quite an odd characteristic, a sort of iconic presence in the foundation of the city (Mumford 1961).

Curiously, and in spite of what is emphatically stated by the historians Plutarch and Diodorus Siculus, the site on which the newly founded town was built was not blessed with any particular qualities of suitability. The city was laid out in a strip enclosed by the sea to the north and west, the marshy lands of the Canopus mouth of the Nile to the east, and the Mareotis Lake to the south, in apparently flagrant disregard of the health and sanitation criteria advocated by Alexander's tutor, Aristotle. Another characteristic which clashes with utilitarian principles is that the orthogonal grid is hardly functional in relation to the landscape, since the longitudinal axes are not parallel to the shoreline. As a consequence, it is natural to seek a possible astronomical orientation (Ferro and Magli 2012). Measuring the azimuth of the Canopic road yields $\sim 65°15'$; the horizon to the east extends towards the Abukir bay and was therefore flat in ancient times; the same is true to the west. In 331 BC this azimuth corresponded to the rising sun on two dates which can be calculated to be around June 2 and July 24 in the Julian proleptic calendar. The second date is very intriguing, since Alexander was born on July 20, a date which is so close as to imply a difference in the azimuth of the rising sun of only 45′.

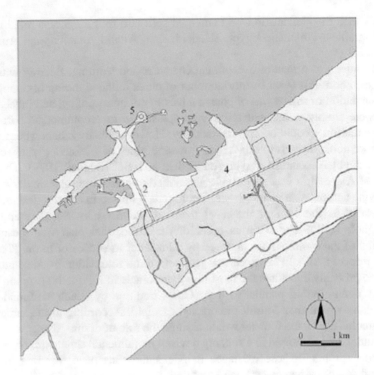

Fig. 10.5 Map of Alexandria at the time of its foundation. *1* Canopic road, *2* Heptastadion, *3* Serapeum, *4* Imperial palace, *5* Isle of Pharos. (*Courtesy* Luisa Ferro)

Moreover, it is well-known that the day of Alexander's birth marked, together with the foundation of the city (Tybi 25 of the Greek calendar, which fell on April 7 in 331 BC) the most important festivity in the town, the celebration of Alexander as a living God. Therefore, it is tempting to conclude that the city was oriented deliberately to the rising sun on the birth date of its founder (Bauval and Hancock 2005; Ferro and Magli 2012). There is, however, a delicate point here because the Julian date of birth of Alexander, of course, has nothing to do with the calendar in use in that period. The planners of Alexandria used the Greek calendar, and we know Alexander was born on the sixth of Hecatombaeon, the first month of the year (Hannah 2005). New Year's Day was the day of the first new moon after the summer solstice; in 356 BC this occurred on July 14 Julian, leading to July 20 for Alexander's birth. This date of course was not a fixed day of the solar year. So, just as in any calendar connected to the moon, feasts like that of Alexander's birth fluctuated through different dates from year to year (in a way similar to what happens with Christian Easter, see e.g. McCluskey 1998, 2006). However, there is no doubt that Alexander's court astronomers were well aware of the cycle of the moon and were able to establish the dates of the new moon in any given year, as the Metonic cycle (meaning that 19 tropical years are needed to complete 235 synodic months) had been known to Greek scientists since the fifth century. Thus the planners of the town could readily rely on this to establish the correct orientation. There is, accordingly, the distinct possibility that the planners of Alexandria really did choose to orient the city astronomically, as a homage to its divine, royal founder. This idea is confirmed by the study of the orientations of other Hellenistic towns of the Seleucid period, such as Seleucia on the Tigris, and by further indications which come from the stars and which are thus quite independent of the calendar (Ferro and Magli 2013). In fact, being oriented to the rising sun at the end of July in the fourth century BC, the city is of course broadly oriented also towards the zodiacal constellation which was hosting the sun at that time: Leo, the Lion, which was therefore the birth sign of Alexander. What is interesting is that, apart from this broad orientation, by reconstructing the sky of the second half of the fourth century BC in Alexandria it can be seen that at that time the most brilliant star of Leo, Regulus, was rising in very good alignment (azimuth 65°20′) with the Canopic road. Regulus had been associated with kingship since Babylonian times, and the Lion image was widely used by the Macedonian kings as a symbol of power. In particular, Alexander's father ordered a huge (more than 6 m tall) stone lion to be sculpted after the battle of Chaeronea, in which he defeated a coalition of Greek city states including Thebes and Athens (Fig. 10.6).

A very similar statue adorned the summit of the Casta hill at Amphipolis, the largest tumulus in the Macedonian age, enclosed in a 500-m-long wall of marble

Fig. 10.6 Keronea. The Lion

and limestone. The tomb inside has only very recently been excavated, and it is evidently a royal tomb, similar in construction and decoration to the tomb attributed to Philip II at Vergina (the burial chamber was still sealed and contained the remains of five different people, probably members of Alexander's family—their identity, however, is wrapped in mystery). Iconography relating to Alexander also featured the lion—even before his birth, as Philip is said to have seen himself, in a dream, securing his wife's womb with a seal engraved with a lion's image. After Alexander, the lion image—now explicitly surmounted by the king's star Regulus —was to appear on coins issued by Cassander, King of Macedon from 305 BC until 297 BC (who was most likely responsible for the executions of Alexander IV and Roxanne) (Fig. 10.7).

To sum up, the image of the Macedonian kings was strongly identified with the lion and it is also likely that the corresponding celestial association was already operational in Alexander's times, although a reconstruction of the history of astrology in this crucial period—when Babylonian, Greek and Egyptian astronomical traditions merge in the cultural environment of Hellenism—is difficult. A further confirmation comes however from the funerary monument of Antiochos I, King of Commagene (Fig. 10.8).

This small kingdom was located between the upper course of the Euphrates and the south-east of Anatolia. Antiochos I reigned here from 69 to 36 BC, cleverly juggling the political interests and influences of the various states in the area, including the Romans. His tomb was built on the summit of the mount called Nemrud Dag and consists of a conical tumulus (which has never been excavated) surrounded by monumental terraces. The east and the west terrace were both embellished with huge limestone statues and reliefs. One such relief—known as the "Lion Horoscope"—depicts Mars, Mercury, Jupiter and the crescent moon in Leo (Neugebauer and Van Hoessen 1959). The two terraces of the monument are

Fig. 10.7 Alexandria, 331 BC. heliacal rising of Regulus

Fig. 10.8 Nemrud Dag. The statues on the eastern terrace, with the (unexcavated) tumulus in the background. The terrace was facing the rising of Regulus at the epoch of construction of the monument

oriented to the solstices, but the huge plinths holding the colossal statues in the eastern terrace point to sunrise on 23 July (and 22 May). The 23 of July appears to coincide with the date of the celebration of Antiochos' ascent to the throne mentioned in the inscriptions of the monument (Belmonte and Gonzalez Garcia 2010). The correspondence with Alexandria is quite striking, considering also that Antiochos makes explicit reference to Alexander the Great in the inscriptions on the monument and that, just as in Alexandria, the terrace also pointed to the rising of Regulus in the years of its construction (Fig. 10.9).

10.3   A Comet and a Capricorn: Augustus' Power from the Stars

It was the evening of June 21, 168 BC, near Pydna, Thessaly. On that day the Roman legions under the command of the consul Lucius Emilius were preparing for the decisive battle of the war against the Macedonian King Perseus. Tension in the camp was at fever pitch: the Roman army was about to fight (and win) against the Macedonian phalanxes, the fearsome battle formations that had enabled Alexander

Fig. 10.9 Nemrud Dag. The stone relief on the western terrace usually called "Lion horoscope". It depicts Mars, Mercury and Jupiter, the crescent moon and the constellation Leo. Regulus is the star pending at the animal's collar

the Great to rule the world not long before. According to Livy, on that evening, the tribune Caius Sulpicius Gallus rallied his troops and informed them that that night, between the second and the fourth hour, the moon would disappear from the sky. He explained that this phenomenon should in no way be taken as a supernatural or inauspicious omen, but as something completely natural, since the Earth sometimes obscures the Moon with her shadow.

From this anecdote we can infer that astronomy in Rome was regularly practised as far back as Republican times; this is backed up also by archaeoastronomical studies (see e.g. Magli et al. 2014). The role of the sky was to become fundamental in Augustus' times, since—as we shall now see—celestial matters played a prominent role in the foundation of the power of the Roman emperor.

The transition from the Roman Republic (*Res Publica*) to government by an absolute ruler was a process which started with Julius Caesar but was brought to completion by his adoptive son Octavian, later called Augustus (23 September 63 BC—19 August 14 AD). Indeed we might say that Augustus' power was born "under a lucky star". The historian Pliny the Elder informs us that (Nat. His. II, 33; translation by the author):

> A comet…appeared on the occasion of the games, that Augustus was celebrating in honour of Venus Genetrix, not long after the death of his father Caesar…the populace attributed to this star the meaning, that the soul of Caesar had been accepted among the immortal gods.

So in 44 BC, a few days after the assassination of Caesar, whilst Augustus was celebrating the games in honour of his deceased (adoptive) father, a comet appeared, so bright as to be visible before sunset. This genuinely took place, as it is confirmed by several other sources and in coinage (Ramsey and Licht 1997). As Livy clearly states, the comet was of great help in constructing the divine nature of the Roman emperor, by means of a mechanism of astral immortality and ascent to the heavens—called *catasterism*—which had very ancient roots (the same mechanism was later to be applied by Hadrian, who deified his favourite, Antinous, claiming that a star appeared in the sky after his death and dedicating to him a constellation).

Thanks to the comet's catasterism, Augustus could thus claim himself as *divi filius*, the son of a God. His consolidation of absolute power started here and went on with the progressive acquisition of *auctoritas*. Auctoritas was a informal construct based—in principle at least—on voluntary consent (Galinsky 2005). For the creation of such consent he drew upon every aspect of Roman public life: civic ideology, traditions, and, of course, religion. Augustus held the post of *pontifex maximus* from 12 BC onward, and even the title of Augustus, bestowed on him by the Senate in 27 BC, had previously been associated only with gods or sacred objects. Augustus thus claimed to be an intermediary with the gods and keeper of the cosmic order, in line with a tradition stemming from the millenarian power of the Egyptian Pharaohs and re-established by the Hellenistic monarchs. The celestial omen of the comet was also taken to symbolise the beginning of a new "golden" age, of which Augustus was, it goes without saying, the natural leader (Zanker 1990).

As we have seen in Sect. 5.3, imposing the cosmic order also means assuring the regularity of the calendrical cycles, and Augustus took this duty very seriously. He was assiduous in completing the Julian reform of the calendar introduced by Caesar. The association between Augustus and the calendar was made tangible by means of a complex and ambitious urban project in a district of Rome called Campus Martius, which had been hitherto reserved for military trials. The Campus Martius was designed by Augustus' son in law Agrippa as a sacred space, where the divine rights of the ruler and his achievements were made concrete as monuments on the earth (Rehak and Younger 2006) (Fig. 10.10).

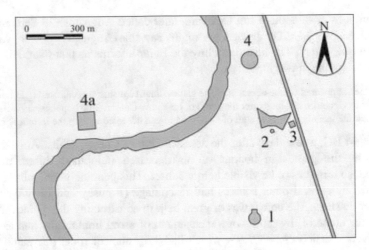

Fig. 10.10 Schematic map of Augustus' monuments in Campus Martius. *1* Pantheon, *2* Horologium Augusti, *3* Ara Pacis, *4* Mausoleum of Augustus, *4a* Mausoleum of Hadrian

The project consisted of the following main elements:

- the mausoleum tomb for Augustus' family, a huge circular building surrounded by trees.
- the Ara Pacis, a marble altar celebrating the ruler and his family
- A Egyptian obelisk, originating in the Sun Temple of Heliopolis. The base of the obelisk (today re-erected, near the original location, in Montecitorio Square) bears an inscription commemorating the settlement of Egypt by Augustus and dedicating the monument to the sun
- *Orologium Augusti*, a sundial which had the above obelisk as gnomon. Of this monument, only (a part of) the meridian line has been recovered so far, under modern buildings, so that it is as yet unknown whether it was used only to indicate noon, or if it was a complete, extended sundial, perhaps connected with a shadow effect to the Ara Pacis (Buchner 1982; Heslin 2007, Frischer 2016).
- The Pantheon, a temple which was subsequently destroyed by fire

In addition, Augustus adopted another powerful celestial icon to legitimise his power: a zodiacal sign. He was born on the 23 September, under the sign of Libra; in spite of this, he chose as the basic leitmotiv for his "power from the stars" another sign: Capricorn.

The Capricorn is an imaginary beast, having the head and front legs of a goat and the tail of a fish. It had already been identified with a zodiacal constellation in Babylonian astronomy, whence it filtered into Greek and then Roman zodiacal iconography and mythology (Fig. 10.11).

The association of Augustus with the Capricorn recurs on innumerable occasions: it is mentioned by several historians (for instance Suetonius), it appears on many coins from various parts of the empire, and is represented in works of art such

Fig. 10.11 Augustus and the Capricorn on a Roman coin (image in the public domain)

as the *Gemma Augustea*. The latter is a veritable masterpiece of fine engraving on ivory, clearly belonging to an high-status personage if not to the imperial family itself. The themes are quite complex, but the centre of focus is Augustus, seated in a hieratic pose similar to that of Jupiter (and typical also of some images of Alexander). A disk containing a star (probably Caesar's comet) with a superimposed Capricorn is floating above him.

There have been many attempts to understand Augustus' choice of Capricorn in astrological terms, for instance, noting that Capricorn occurs 9 months before the date of birth and so could be considered as the sign denoting his conception, or analysing the position of the Moon in Augustus' horoscope (Barton 1995; Hannah 2005). In fact, we do not know what strategy was dreamt up by the ruler's astrologers to legitimise Capricorn, but what is certain is that astrology was—already at that time—a flexible tool and Augustus exploited it for his personal propaganda. So, the Libra-born Augustus chose Capricorn for utilitarian reasons, the main one probably being that it was the sign hosting the winter solstice (as a matter of fact, due to precession, the winter solstice had just left the *formal* boundary of Capricorn in favour of Sagittarius, but this is clearly just a technicality, as Capricorn is *still today* the sign of those born at the winter solstice—for those who believe in astrology, that is).

As we have seen, the winter solstice was construed in countless cultures as a symbol of rebirth, and therefore it comes as no surprise that Augustus' new era iconography was linked with such a celestial icon; what is more, Capricorn was associated with fortune and was sometimes represented by a cornucopia.

This Capricorn iconography was widely disseminated throughout the empire by various means, and was also adopted in the emblems of some legions. Interestingly, the ruler's association with the winter solstice appears to have influenced also the planning of newly founded towns. This is particularly apparent at *Augusta Praetoria Salassorum*, today's Aosta, in northern Italy (Fig. 10.12).

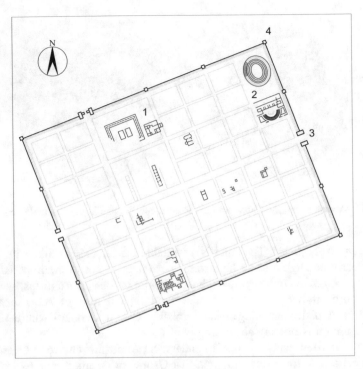

Fig. 10.12 Schematic map of Augusta Praetoria Salassorum, the Roman Aosta. *1* Twin temples and forum, *2* Amphitheatre and theatre, *3* Porta Praetoria, *4* Balivi tower ▶

The town was founded by the Roman Army in 25 BC shortly after the war against the Salassi, the original inhabitants of the region, who blocked the consular road leading to Gallia (Gaul). The foundation of Aosta therefore came out of practical considerations—to make the presence of the Romans in the area permanent—as well as for symbolic reasons, with the aim of marking the place as a sacred landscape of victory (Laurence et al. 2011). The city lies at some 600 m above sea level, with prominent mountains that dominate the landscape. In particular, the horizon to the east is occupied by the 3000-m-high Emilius ridge, whose peak is located at a distance of less than 9 km as the crow flies. As was standard in Roman colonies, the plan of the city is rectangular, and the blocks are arranged on a orthogonal grid, with two main urban axes (usually called Kardo and Decumanus in present day literature, although this terminology tended to be used by the Romans for *centuriations*, the agricultural divisions of land) dictating the urban plan. Luckily, much remains of the Roman town; we have most of the walls and one of the four main gates, the eastern one or Porta Praetoria; near Porta Praetoria lies the equally famous theatre, with its 22 m-high façade. Another monument, almost intact, is a huge triumphal arch devoted to the ideal founder, Augustus. The layout of the town roughly follows the topography of the valley, but there was no compelling constraint determining the specific orientation of the axes. Their azimuths, measured with a precision theodolite, are 68°02′ and 158°06′ (so the ancient surveyors were extremely precise in maintaining orthogonality). Taking into account the height of the horizon, it can be seen that the sun rises along the first axis in the days around June 3/July 10. In Italy the solar amplitude is much less than 90° so in principle one would not expect a solar orientation of the second axis; however, this situation may change drastically if the horizon is very high. This is precisely the case in Aosta: the midwinter sun rises with a theoretical azimuth of 125° but remains behind the Emilius ridge until it reaches an altitude of up to 17°. The azimuth of the sun at this altitude turns out to be very close to that of the Kardo at 158°: Aosta is thus a city oriented to the winter solstice sunrise (Bertarione and Magli 2015) (Figs. 10.13 and 10.14).

Fig. 10.13 Aosta, winter solstice. The sun rises in alignment with the "Kardo" of the town. (*Courtesy* Paolo Pellissier)

Fig. 10.14 Aosta, 1 century BC. The winter solstice sun rising between Capricorn and Sagittarius at the time of foundation of the city

The case of Aosta is not unique, as there are many Roman towns oriented to the sun rising or setting at a solstice; some examples are Norba, dated to the second half of the second century BC, Metellinum, founded in 79 BC, Verona, with its urban plan of probably Julian foundation, and Liberalitas Iulia, Augusta Raurica and Emerita Augusta, all founded under Augustus rule (Magli 2008; González García et al. 2014).

The intention of the builders to align the city towards the sun rising at winter solstice, and therefore towards Capricorn, the sign of Augustus, is confirmed by the recent discovery of a sculpted block located on the north-east corner tower (Bertarione 2013). The reliefs on the blocks allude to the ceremony of foundation of the colony, which was a religious rite meant as a sort of ideal replication of the mythical foundation of Rome. The ritual was quite elaborate: the whole (sacred, inviolable) perimeter of the future city walls was ploughed by pairs of yoked oxen ridden by priests and assisted by functionaries, and the plough was raised at the future gates to allow, in the future, people to pass without breaking the sacred furrow (Rykwert 1999). As a consequence of this ritual, the reliefs include—besides two phallic symbols, with an apotropaic function, to ward off evils and afford magical protection for the newly founded town—the hand-drive of a plough and a arrow-like ploughshare. The figure of a rampant animal is also visible, which —although much weathered—may feasibly be interpreted as a Capricorn.

To conclude then, archaeoastronomy shows us that Aosta, as was the case with Alexandria three centuries before, was born under her founder's (adoptive) sign.

10.4 Astronomy and Empire at the Pantheon

The Roman Emperor Hadrian was born on 24 January 76 AD in Spain. He was the son of one of the cousins of Trajan, who appointed him as his successor shortly before dying. The new emperor was blessed with a profound cultural curiosity which, combined with the official and military duties, prompted him to visit almost every part of the empire. Furthermore, Hadrian had inherited from his predecessor —who commissioned the construction of his own forum and created that stunning specimen of Roman marble carving, Trajan's column—a passion for architecture and the desire to ennoble Rome and her empire with architectural masterpieces. In particular, he commissioned the splendid Villa Adriana at Tivoli, royal retreat and estate (the size of a small town) and he refurbished Campus Martius where he re-built the Augustan Pantheon as an incomparable masterpiece of dome architecture.

Hadrian's Pantheon is the best preserved monument of the imperial period in Rome (Fig. 10.15). It stands over the foundations of the temple originally built under Augustus' rule, which had been destroyed by fire; Hadrian's architects completed the reconstruction in AD 128 (Wilson Jones 2003). As a symbol of continuity, the original dedicatory inscription by Agrippa was affixed to the new façade, where it can still be read. The building consists of a rectangular portico, with three lines of granite columns, fronting a circular building. The latter is a huge hemispherical dome (43.3 m in diameter), built over a cylinder which has the same diameter and is as high as the radius. Therefore, the ideal completion of the upper hemisphere by a hypothetical lower one touches the central point of the floor, directly beneath the only source of natural light of the building. This source of light is the so-called *oculus*, a circular opening 8.3 m wide on the top of the cupola. It is the only source of direct light since no direct sunlight can enter from the door throughout the year, owing to the northward orientation of the entrance, which points $\sim 5°30'$ west of north. A curious consequence of this orientation is that the huge mass of the building gives visitors an eerie impression of cold and dark, quite an atypical feature when compared with the temples of the classical age (think, for example, the Greek temples of Sicily, Sect. 10.1). We shall see in a moment how this impression of coldness is turned on its head on certain days of the year, by means of a spectacular hierophany. Of the original embellishments the interior of the building would have had, the coffered ceiling, part of the marble interiors, the bronze grille over the entrance and the great bronze doors have survived. The interior wall, although circular in plan, is arranged into sixteen regularly spaced sectors: the northernmost one contains the entrance door, and then (proceeding clockwise) pedimented niches and columned recesses alternate, which were probably intended for statues.

The Pantheon has exerted a tremendous influence on architecture since the Renaissance. Yet in spite of having such a prominent role in history, we know very little about its function, as only two Roman sources mention it: Pliny, who was

Fig. 10.15 Rome. The pantheon, front view ⊙

writing, however, before Hadrian's reconstruction, and the historian Cassius Dio, writing some 70 years after Hadrian, who made the following cryptic statement:

> Perhaps it has this name because, among the statues which embellished it, there were those of many gods, including Mars and Venus; but my own opinion on the origin of the name is that, because of its vaulted roof, it actually resembles the heavens. (Cassius Dio 53.27.2)

So we do not really know why the Pantheon was built or how it was used: the Pantheon is a one-of-its-kind masterpiece, but its builders left nothing in writing. One thing, however, is certain: on each sunny day, the attention of the visitors to the Pantheon is attracted by the shaft of sunlight from the oculus that penetrates the gloom. As a matter of fact, if a systematic analysis is applied to the motion of the sunbeam inside the monument, it is possible to show that it is precisely this movement that is the key to the whole project, since it governed the entire design from the very beginning (Del Monte 1990; Hannah and Magli 2011) (Fig. 10.16).

The general idea was probably inspired by a particular type of Roman sundial, which captured the beam of sunlight within a shadowy interior (Hannah 2009). This device consisted of a stone block carved out into a hollow hemisphere, with a hole left in its upper surface, through which the sun filtered on to the graduated surface

Fig. 10.16 Rome. The Oculus in the pantheon

inside. Clearly, for this type of sundial to work correctly, the stone face had to be oriented to due south, while the hollow hemisphere faced north. The formal similarity with the project of the Pantheon is thus self-evident, although, of course, the building should not be construed as any sort of precision instrument. As will be clear shortly, its connection with the cycle of the sun was not intended for astronomical observation but rather to highlight the relationship between Roman power and the cosmos.

The main features of the sun-based project of the Pantheon can be understood by keeping the time at (local) noon fixed over the course of the year and studying the position of the sunbeam day by day at this specific time. Since the door opens to the north, the shaft of sunlight at noon is always located on an approximate meridian line which passes over the entrance. At the autumn equinox, the spot of sunlight touches the interior springing of the upper hemisphere. Then the spot moves up to its maximum height in the roof over the entrance, reached at winter solstice. Thereafter, it starts moving down, again brushing the base of the dome at the spring equinox. In the subsequent days, the sun culminates progressively higher, and so the beam moves further down, illuminating the grill over the entrance, and finally the entire opening, which is fully lit around 21 April. After that, the beam starts moving across the floor towards the centre of the building (which is never reached since, of course, the sun does not cross the zenith at the latitude of Rome). From the summer solstice the beam "turns back", re-crossing the entrance between the end of August and the autumn equinox.

We can conclude, accordingly, that the dimension of the oculus was fixed in such a way that the angle between the spring at the base of the dome and the external rim of the oculus coincides with the altitude of the sun in the days in which the springing had to be illuminated, namely the equinoxes. In this way the sun "spends" autumn and winter in the upper hemisphere of the building but, immediately after the March equinox, begins to be visible *from outside,* due to the presence of the grille. This effect increases gradually up to the 21 April, when the maximum illumination of the entrance is attained (a phenomenon probably already noticed by Francesco Piranesi, who depicted it in one of his etchings in the eighteenth century) (Fig. 10.17).

We can, therefore, say that the whole architecture of the Pantheon was conceived of to attract attention to the equinoxes and to the 21st of April. Why? As we have seen in the previous section, Roman religion underwent a reassessment, aimed at accommodating the divine nature of the emperor precisely during the years of the first building of the Pantheon, under Augustus. The emperors wished to promote an association of their power with the sky and there are passages from Latin authors which suggest that a particular point of cosmic balance was understood to be assigned to the emperor in the heavens. Virgil for instance (*Georgics* 1. 24–35) places Caesar between Virgo and Scorpius, so in the equinoctial sign Libra. Manilius (*Astronomica* 4. 546–551, 773–777), writing in Tiberius's reign, again places the emperor in Libra,

Fig. 10.17 Rome. The interior of the Pantheon as depicted by Francesco Piranesi in the eighteenth century (image in the public domain)

as a point of balance between the ecliptic and the equator. Lucan (1.45–59), writing in Nero's time, has the apotheosised emperor joining the heavens and finding his proper seat on the celestial equator, where he will ensure balance and stability (apparently, Nero took such ideas very seriously, and indeed in his enormous palace, the Domus Aurea, a sort of forerunner of Hadrian's Pantheon can be seen, in the form of a majestic octagonal hall where a equinoctial hierophany manifests itself; see Oudet 1992; Voisin 1987; Hannah et al. 2015).

So far, so good, as far as the equinoxes are concerned. But what about the 21st of April? The month of April was traditionally devoted to Venus, the goddess from whom the Caesar family Gens Julia claimed direct lineage, and the 21st of April is the traditional date of the foundation of Rome (see, e.g., Ovid, *Fasti* 4: 721–862). Therefore, the symbolic action of the sun on this day is "to put Rome among the Gods". If we suppose, as seems likely, that the emperor was celebrating this very day at the Pantheon, then his entrance together along the sun at noon would have been a symbolic link between the people and the gods, as well as an impressive confirmation of the emperor's power and divine nature (Fig. 10.18).

So the Pantheon, besides being an unparalleled engineering masterpiece, was also a cosmological signpost, an icon ideally linking sun and time to the cosmic order.

Perhaps, the most appropriate place to conclude our long journey devoted to the history of the relationships between astronomy, architecture, and power.

Fig. 10.18 Rome, 21 April. The light from the Oculus of the Pantheon hits the entrance at local noon ▶. (*Courtesy* Francesca Agostino)

References

Aveni, A. F., & Romano, G. (2000). Temple orientation in Magna Grecia and Sicily. *Journal for the History of Astronomy, 31,* 52–57.

Bagnall, R. S. (1979). The date of the foundation of Alexandria. *American Journal of Ancient History, IV,* 46–49.

Barton, T. (1995). Augustus and Capricorn: Astrological polyvalency and imperial rhetoric. *The Journal of Roman Studies, 85,* 33–51.

Bauval, R., & Hancock, G. (2005). *Talisman.* Toronto: Anchor Canada.

Belmonte, J. A., & Gonzalez Garcia, C. (2010). Antiochos's hierothesion at Nemrud Dag revisited: Adjusting the date in the light of astronomical evidence. *Journal for the History of Astronomy, 41,* 469–481.

Bertarione, S. (2013). Indagini archeologiche alla Torre dei Balivi. In *Si svela la sanctitas murorum* (pp. 9). BSBAC.

Bertarione, S., & Magli, G. (2015). Augustus' power from the stars and the foundation of Augusta Praetoria Salassorum. *Cambridge Archaeological Journal, 25*(2015), 1–15.

Boutsikas, E. (2009). Placing Greek temples: An archaeoastronomical study of the orientation of ancient Greek religious structures. *Archaeoastronomy, 21,* 4–19.

Boutsikas, E., & Ruggles, C. L. N. (2011). Temples, stars, and ritual landscapes: The potential for archaeoastronomy in ancient Greece. *American Journal of Archaeology, 115*(1), 55–68.

Buchner, E. (1982). *Die Sonnenuhr des Augustus.* Mainz am Rhein: von Zabern.

Castagnoli, F. (1971). *Orthogonal town planning in antiquity.* Cambridge, MA: MIT Press.

Del Monte, C., & Lanciano, N. (1990). L'occhio di luce: il Pantheon. Il Manifesto, 22 July.

Dinsmoor, W. B. (1939). Archaeology and astronomy. *Proceedings of the American Philosophical Society, 80*(1), 95–173.

Ferro, L., & Magli, G. (2012). The astronomical orientation of the urban plan of Alexandria. *Oxford Journal of Archaeology, 31,* 381–389.

Ferro, L., & Magli, G. (2013). Astronomy and perspective in the cities founded by Alexander the Great. *Aplimat, 5,* 523–532.

Frischer, B. (2016). Edmund Buchner's Solarium Augusti—New observations and simpirical studies. *Atti Della Pontificia Accademia Romana di Archeologia, LXXXIX,* 3–73.

Galinsky, K. (2005). *The Cambridge companion to the age of Augustus.* Cambridge: Cambridge University Press.

González García, C., Rodríguez Antón, A., & Belmonte, J. A. (2014) On the orientation of Roman cities in Hispania: Preliminary results. *Mediterranean Archaeology and Archaeometry, 14*(3), 107–119.

Hannah, R. (2005). *Greek and Roman calendars: Constructions of time in the classical world.* London: Duckworth.

Hannah, R. (2009). *Time in antiquity.* Routledge: London.

Hannah, R. (2013). Greek temple orientation: The case of the older Parthenon in Athens. *Nexus Network Journal, 15,* 423–443.

Hannah, R., & Magli, G. (2011). The role of the sun in the Pantheon design and meaning. *Numen, 58,* 486–513.

Hannah, R., Magli, G., Palmieri, A. (2015). Nero's "solar" kingship and the architecture of Domus Aurea. *Numen, 63,* 511–524.

Hannah, R., Magli, G., Orlando, A. (2017). Astronomy, topography and landscape at Akragas' valley of the temples. *Journal of Cultural Heritage, 18.*

Hannah, R., Magli, G., & Orlando, A. (2018). The role of urban topography in the orientation of greek temples: The cases of Akragas and Selinunte. *Mediterranean Archaeology and Archaeometry, 16,* 213–217.

Heslin, P. (2007). Augustus, domitian and the so-called horologium augusti. *Journal of Roman Studies, 97,* 1–20.

Lane Fox, R. (2004). *Alexander the Great.* London: Penguin Books.

Laurence, R., Cleary, S. E., & Sears, G. (2011). *The city in the Roman West*. Cambridge: Cambridge University Press.

Lawrence, A. W. (1996). *Greek architecture*. Yale University Press.

Liritzis, I., & Castro, B. (2013). Delphi and cosmovision: Apollo's absence at the land of the hyperboreans and the time for consulting the oracle. *Journal of Astronomical History and Heritage, 16*, 184–206.

Liritzis, I., & Vassiliou, H. (2003). Archaeoastronomical orientation of seven significant ancient Hellenic temples. *Archaeoastronomy: The Journal of Astronomy in Culture, 17*, 94–100.

Magli, G. (2008). On the orientation of Roman towns in Italy. *Oxford Journal of Archaeology, 27* (1), 63–71.

Magli, G., Sampietro, D., Realini, E., & Reguzzoni, M. (2014). Uncovering a masterpiece of Roman architecture: The project of Via Appia between Collepardo and Terracina. *Journal of Cultural Heritage, 15*, 665–669.

McCluskey, S. C. (1998). *Astronomies and cultures in early medieval Europe*. Cambridge University Press.

McCluskey, S. C. (2006). The orientations of medieval churches: A methodological case study. In T. W. Bostwick, B. Bates (Eds.), *Viewing the sky through past and present cultures. Pueblo Grande Museum anthropological papers no 15* (pp. 409–420). Phoenix: City of Phoenix.

Mumford, L. (1961). *The City in history*. London: Secker and Warburg.

Neugebauer, O., & Van Hoessen, H. (1959). Greek horoscopes. *Memoirs of the American Philosophical Society, 48*, 14–16.

Nissen, H. (1873). Über Tempel-orientierung. *Rheinsiches Museum für Philologie, 28*, 513–557.

Oudet, J. -F. (1992). Le Panthéon de Rome à la lumière de l'equinoxe. In *Readings in archaeoastronomy. Papers presented at the international conference: Current problems and future of archaeoastronomy held at the State archaeological museum in Warsaw, 15–16 November 1990* (pp. 25–52). Warsaw: State Archaeological Museum.

Penrose, F. C. (1897). On the orientation of Greek Temples and the dates of their foundation derived from Astronomical considerations, being a supplement to a paper published in the 'Transactions of the Royal Society,' in 1893. In *Proceedings of the royal society of London* (Vol. 61, pp. 76–78).

Ramsey, J. T., & Lewis Licht, A. (1997). *The comet of 44 B.C. and Caesar's funeral games*. Atlanta: Scholars Press.

Rehak, P., & Younger, J. C. (2006). *Imperium and cosmos: Augustus and the northern Campus Martius*. Madison: University of Wisconsin Press.

Retallack, G. (2008). Rocks, views, soils and plants at the temples of ancient Greece. *Antiquity, 82* (317), 640–657.

Rykwert, J. (1999). *The idea of a town: The anthropology of Urban Form in Rome, Italy, and the ancient world*. Cambridge: MIT Press.

Salt, A. (2009). The astronomical orientation of ancient Greek Temples. *PLoS ONE, 4*(11), e7903.

Salt, A. M., & Boutsikas, E. (2005). Knowing when to consult the Oracle at Delphi. *Antiquity, 79* (305), 564–572.

Scully, V. J. (1962). *The earth, the temple, and the gods: Greek sacred architecture*. New Haven: Yale University Press.

Voisin, J. -L. (1987). Exoriente Sole (Suétone, Ner. 6). D'Alexandrie à la Domus Aurea In L'Urbs: Espace urbain et Histoire (pp. 509–41). Rome: École française de Rome.

Wilson Jones, M. (2003). *Principles of roman architecture*. New Haven and London: Yale University Press.

Zanker, P. (1990). The power of images in the age of Augustus. The University of Michigan Press.

Chapter 11
Asian Cultures

11.1 The Terracotta Army

The history of China is classified according to dynasties, starting from the second millennium BC. The history of imperial China formally begins, however, from the reign of the first emperor of Qin, Shi Huang, who succeeded in unifying the country in 221 BC. His name is famous worldwide due to his tomb, located in Lintong to the south-east of Xi'an, and to an astonishing archaeological discovery occurred there: the so-called Terracotta Army. It is a collection of thousands of terracotta statues of warriors (Fig. 11.1). Overall, there are more than 8000 soldiers, together with hundreds of chariots and horses, located in 3 huge underground pits. The statues are life-sized and made out choosing among several different possible faces, hairstyles, uniforms and so on, so that they practically look all different from each other. They were originally painted with bright colors, most of which have disappeared.

The warriors are disposed in rows, ready for a battle or—at least in the view of many, including myself—for an official ceremony to which the Emperor must attend in the afterworld. There are indeed not only warriors' pits: other excavations have revealed a rich—and sometimes puzzling—symbolic equipment for the emperor's afterlife. First of all, a pit containing two half-sized bronze chariots, with all probability representing the emperor's official procession. Further, a rectangular pit was found containing thousands of stone pieces which, when assembled, reveal to form stone (and therefore purely symbolic) armors, perhaps meant for the warriors in case of needs against the spirits. Yet another pit contains terracotta statues of acrobats and musicians. Last but not least, a small bronze lake surrounded by bronze water birds of different species has also been found (Zhewen 1993; Wu 2010).

The tomb lies beneath a huge burial mound of rammed earth, which works as an unmistakable landmark denoting the funerary landscape of the Emperor. The burial chambers have not been excavated, but the Chinese historian Sima Qian describes

© Springer Nature Switzerland AG 2020
G. Magli, *Archaeoastronomy*, Undergraduate Lecture Notes in Physics,
https://doi.org/10.1007/978-3-030-45147-9_11

Fig. 11.1 The terracotta warriors, Pit. n.1

them as a microcosm endowed with vaults representing the heavenly bodies and a miniature of the empire—including rivers made out of mercury—on the ground.

The tomb was positioned near the sacred peak of Mt. Li, which dominated the Wei River from the south. The mound was itself referred to as a "mountain", and was thus meant to be a replica of nature, over which the owner exerted his power and control. The base plan of the mound is however square and oriented to the cardinal points, as well as the pits of the terracotta warriors. We thus see in the project the symbolic importance of cardinal orientation on Earth, connected as it was with the "cosmic" order in the heavens, which formed the core organizing principles of the Chinese doctrine of power: the Mandate of Heaven. The emperor indeed conceived of himself as the earthly representative of the tutelary gods of heaven and thus was pivotal in the proper functioning of the earth and its inhabitants. In particular, astronomy and power were deeply and intimately connected, and the function of the celestial pole as pivot of the sky was equated with the centrality of the imperial power on Earth. As a result of the celestial identification of the Emperor, the whole polar region of the sky was identified as a celestial image of the Emperor's palace, the "Purple Enclosure", bounded by a circle of about 15°, and many stars of the enclosure received a name symbolically related to this identification (Needham 1959; Didier 2009; Pankenier 2009, 2015).

11.2   The Pyramids of Ancient China

The realm of the first Emperor rapidly became unstable after his death, but China was unified again very soon under Liu Bang, the founder of the Western Han Dynasty. This Dynasty (202 BC–9 AD) marked important political, economic and scientific developments. Under the Han, for example, a first version of the magnetic compass was probably invented (Needham 1970). The Han rulers followed Qin's custom of being buried in tombs located under huge, square-base mounds; their shape and aspect closely recalls pyramids, and indeed they are known popularly as "Chinese pyramids".

These monuments make for a fascinating presence in the rapidly developing area of Xian; the hugest one, Maoling of Emperor Wu, has a square base with sides of almost 250 m (Fig. 11.2). The mausoleums are mostly located along the vast flatland to the north of the river Wei; altogether, they create a sacred landscape, dotted over the course of the river for some 40 km. Their spatial distribution is interesting: as matter of fact, they were not built in a linear succession from east to west along the river, but what seem to be "jumps" back and forth occurred. This probably corresponds to a doctrine called Zhaomu—mentioned in historical sources —that states that left/right (east/west) have to be alternately selected for emperor's burials, so that, looking at a tomb, the first successive one will be found to the left (west) and the second one to the right (east).

Further, a series of visual, topographical relationships connect these monuments trough "dynastic" lines of sight, in a way similar to the visual axes of the Egyptian's

Fig. 11.2   The huge mound of the Maoling mausoleum, covered by trees ▶

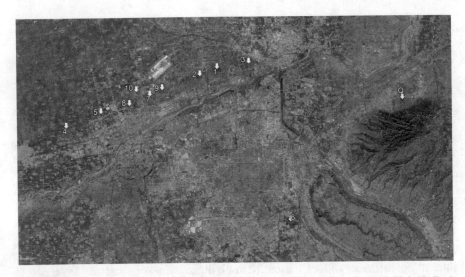

Fig. 11.3 The pyramids of the Western Han emperors. *1* Gauzu (Chanling), *2* Hui (Anling), *3* Jing (Yanling), *4* Wu (Maoling), *5* Zhao (Pingling), *6* Xuan (Duling), *7* Yuan (Weiling), *8* Cheng (Yangling), *9* Ai (Yiling), *10* Ping (Kanling). The mausoleum of the first Emperor of Qin is denoted by Q. (Image courtesy Google Earth, editing by the author)

pyramid fields (Figs. 11.3 and 11.4). For example, the centers of the tombs of Emperor Yuan, of his son Cheng and of his grandson Ai are connected by a almost perfect straight line. Another stunning example is the placement of the pyramid of Emperor Hui with respect to those of his parents. His father, Emperor Gaozu, and his mother, Empress Lu, are indeed buried in two huge, twin mausoleums which form a "sign of two mountains" at the eastern horizon of Hui's tomb, whose north-west side is aligned, to a distance of about 4 kms, with the south-east side of Empress Lu tomb.

Like that of Qin, also the tombs of the Han emperors are unexcavated, but a series of burial pits has been discovered near one of them, Emperor Jing's mausoleum. They contain an enormous quantity of terracotta statues. They are, however, miniatures, contrary to Qin life-sized warriors.

The orientations of the Han monuments divide into two "families" (Magli 2018, 2019a, b). Family 1 comprises mounds whose square base is precisely oriented the cardinal points, with errors not exceeding ±1°, while Family 2 comprises mounds with a rough orientation to the cardinal points, but with errors in relation to the geographic north of several degrees. However, the errors are not randomly distributed: they are always to the west of north and exhibit a tendency to decrease in time from a maximum of 14° to a minimum of 8°.

The mounds of Family (1) were certainly oriented by determining the cardinal directions. The orientations of family (2) cannot be explained as arising from errors in determining true north, both because the errors would be too large and because they are not randomly distributed. A likely solution is that they are oriented to

Fig. 11.4 A picture taken along the north side of the pyramid of Emperor Hui of Han. In the foreground (left), the huge mole of one of his satellites, the pyramid of Marquis Zang. At the horizon, a "sign of two pyramids", (outlined) those of the Emperor's parents Gaozu and Lou ⊙

another target: not to the north celestial pole but to the maximal western elongation of a circumpolar star with respect to the pole—and therefore to an azimuth equal to the distance in degrees of the very same star to the pole. To understand which star was chosen, we must first of all remember that, due to precession, the north celestial pole was in a dark region. In fact, the pole had not been located sufficiently close to any bright "polar" star since the third millennium BC, when the pole star was Thuban, of the constellation Draco. Later the pole had been relatively close to the bright star Kochab, but in Han times it had already started to move towards "our" pole star, Polaris. The target could therefore be one of these two stars. Polaris seems preferable because, if we follow its slow apparent approach to the pole due to precession, we see that its maximal western elongation decreases in (rough) agreement with the gradual shift in the orientations of the Han mounds of Family (2), being $\sim 13°$ in 200 BC, $12° 20'$ in 100 BC, and $11° 40'$ in 1 AD.

To explain the symbolic meaning of all these orientations, we must consider that the polar region and the North celestial pole remained tightly associated with the emperor during the whole Chinese history. The Han rulers were thus, in principle, meant to be located in the north, looking south over the kingdom—thus explaining the orientations of Family (1). Further, as mentioned above, the whole polar region of the sky was identified as a celestial image of the Emperor's palace. The Chinese astronomers very probably knew (by direct observation) that, inside the Purple enclosure, Kochab had already reached the minimal distance from the pole (this occurred around the sixth century BC) and thus that the pole, the center of the Purple enclosure, was moving away from Kochab to Polaris. As far as the choice of the of the "hand" of the elongation measured (either east or west with respect to true north) is concerned, one possible explanation is seasonality, that is, that the orientation ceremonies were all made in the same period of the year.

After the decline of the Han dynasty, the construction of "pyramids" in China comes to the end. The great emperors of the Tang dynasty (618–907 AD) built their tombs under already existing mountains, endowing them with spectacular "sacred ways": paved ceremonial roads flanked by dozens of monolithic statues and endowed with huge monolithic stelae. The most spectacular of these tombs is the Qianling Mausoleum, the tomb of emperor Gaozong (649–683 AD) and of his wife Wu Zetian located in Qian County, some 80 km northwest of Xi'an (Fig. 11.5).

During the Tang, the traditional Chinese doctrine called Feng Shui became progressively relevant (Bruun 2008). The key idea of Feng Shui is that the earth is criss-crossed by the so-called Qi, the "vital energy", and that auspicious places for benefiting of the flow of this energy can be found by analyzing the morphology of the terrain and/or the directions of the magnetic compass. Of course, these ideas have no scientific basis; as a matter of fact however, Feng Shui landscapes are based on a conception of ideal beauty which inspired spectacular sacred landscapes such as those of the Ming and of the Qing royal necropolises (Paludan 1981, 1991). Essentially, a auspicious place must have a mountain to the north, undulating "dragon" hills to east and west, and flowing of water to the south. Being criss-crossed by topographical alignments—for instance, between tombs, sacred

Fig. 11.5 The Qianling mausoleum

ways, and mountains—these landscapes can be efficaciously studied with the same methods that Archaeoastronomy employs for astronomical alignments (Magli 2019a, b).

11.3 Between Sun and Waters: Angkor Wat and the Khmer State-Temples

The Khmer empire flourished between the 8th and the fourteenth century AD in the Cambodian lowlands. The Khmer kings adopted the construction of gigantic temples as a means for the explicit representation of their power; as a consequence, a series of masterpieces—the so called "state temples"—were constructed. Among the Khmer state temples, Angkor Wat singles out in being one of the hugest temples ever built on planet earth (Jacques and Lafond 2004).

Geographically, most of the state temples concentrate in the Angkor area, today's Siem Reap (there are, however, two main exceptions: Koh Ker, located in northern Cambodia 85 km north-east of Angkor, and Preah Khan of Kompong Svay, 100 kms to the east). The temples are vast rectangular enclosures, with a central unit and several auxiliary buildings and shrines. Generally speaking, it can be said that they functioned as royal residence and as main center of cult attesting to the beliefs and religiosity—either Buddhist or Hinduist—of the king. Symbolically, these constructions legitimated the ruler's power, attesting at his divine nature. Up to a few years ago the Khmer temples were even conceived as "concentrated state towns" but recent research and mapping on large scale has shown the complexity of the urbanization of the whole Angkor area, putting aside the idea of the state-temples as "capital cities" (Fletcher et al 2015). A further, funerary function of the temples as tombs for the kings, although likely, has never been proved, and the burial place of the Khmer kings remains unknown.

Together with the temples, many huge *barays* were built. The barays are water reservoirs of strictly rectangular form, oriented to the cardinal points. The most imposing one, the West Baray, is about 8 by 2 km. With a capacity which exceeds by far 50 millions of cubic meters of water, it is still today the largest fully-artificial lake in the world. As a consequence of the construction of these hydraulic systems, the landscape of Angkor is fascinatingly suspended between sky and water. The barays thus had both a practical function—acting as central collectors of water for agricultural distribution—and a ritualistic function, in being explicitly associated with the construction of the state temples. To confirm the ritualistic function, in the center of the barays other temples—called Mebons—were built, constructing to this end huge artificial islands.

Since the state temples reflected concepts related to the foundation of power and to the cosmic order, cosmological symbolism is self-evident in all these buildings; in particular, the characteristic layout of many temples—a "pyramid mountain" surrounded by a moat—is usually considered to correspond with the "landscape" of the Indian cosmological myth, where a mountain (Mount Meru) is surrounded by a sea of milk from which ambrosia is churned by gods and demons.

Religious symbolism associated with cosmic order is, not surprisingly, reflected in the orientations of the temples which are, generally speaking, cardinal. This holds, in particular, for the place from which we shall start our visit: Angkor Wat. Originally built as an Hinduist temple by king Suryavarman the second at the beginning of the twelfth century AD, Angkor Wat's main entrance is from the west, probably because it was devoted to Vishnu, a God tightly connected to this cardinal direction. The dimensions of Angkor Wat are striking: it is enclosed by a moat more than 5 km long and by a perimeter wall of 3.6 km. The core of the temple is the so called quincunx. A quincunx is an ensable of four towers with a fifth, higher one at the center (the central tower at Angkor Wat raises as high as 63 m). The main entrance is accessed by a stone bridge guarded by auspicious animals: the lion and the Naga, a serpent with seven heads. The temple proper stands on a raised terrace and is composed of three levels of rectangular galleries enclosing the central tower. The galleries are decorated with bas-reliefs of mythological and historical content. The most famous relief is certainly that representing the churning of the Sea of Milk: Vishnu is in the center, while asuras—demons—and devas—gods—use the giant serpent Vasuki as churner.

In Cambodia, the climate is tropical and the arrival of the rainy season in middle Spring is the fundamental climatic event. Not surprisingly then, the Angkor Wat temple was connected to the celestial cycles trough a spectacular hierophany which occurs at the Spring equinox, the day which is the harbinger of the rainy season (of course, the phenomenon occurs also at the autumn equinox). To understand how this hierophany works, one should notice that the azimuth of the axis from outside looking towards the temple is 90.5°. This slight deviation from the true east has an interesting consequence. Looking from the main, west gate at dawn at the equinoxes, the sun is seen to rise just above the central tower, "crowning" it almost vertically. The reason is that at the latitude of Angkor the trajectory of the sun is very steep, and therefore a small increase in azimuth leads to a strong increase in height; the "horizon height" of the central tower of Angkor Vat from the western entrance is about 5° and the center of the sun at the equinoxes reaches such an altitude at an azimuth of 90° 40′ (Fig. 11.6).

What about the meaning of such spectacular hierophany? Many "esoteric" interpretations of the temple exists, according to which it should be a sort of huge calendrical device enclosing a complex series of further astronomical alignments and incorporating "astronomical numbers" in its dimensions (see e.g. Stencel et al. 1976). However, this "numerology" has no historical basis and the idea that the temple was used as an astronomical device has nothing to do with the way of thinking of the builders: the beautiful hierophany of the sun suspended just above the mountain-temple at the equinox was rather intended as a symbolic materialization of the connection of the temple itself with the heavens and, as a direct consequence, of the divine rights of the ruler.

Besides Angkor Wat—whose main entrance is to the west—the azimuths of all the other main temples (from outside looking along the axis from the main

Fig. 11.6 Angkor Wat at dawn at the Spring equinox: the sun crowns the central tower ▶

entrance) lie within an interval as small as five degrees to the south of west (Magli 2017a, b). As in Angkor Wat, these slight deviations from the east-west line are probably intentional. It is indeed clear from the monuments themselves that the Khmer architects were extremely precise; further, if the observed deviations were originated by errors of measure, then either a method to find the cardinal directions which led only to south of west errors was devised (a thing which looks unlikely) or the results should distribute randomly on both sides of the expected value of 270°. The key to this riddle is to notice that, if a phenomenon similar to that occurring at Angkor Wat has to be observed in a temple whose main access is to the east, it must occur at sunset, and the azimuth of the temple must be slightly misaligned to the south of west. In fact in this way the equinoctial sun will be seen to crown the central tower and then disappear "inside" the temple immediately before sunset. The phenomenon becomes striking at the second world-wide famous Angkor temple, the Bayon.

The Bayon is a Buddhist temple, constructed around the end of the twelfth century AD by King Jayavarman VII. It is characterized by the ubiquitous presence of "smiling" stone faces distributed on all his towers. The faces probably represent the king as a manifestation of Lokesvara—a compassionate person who is able to reach nirvana but delays doing so to help suffering people. The azimuth of the axis of the temple from outside looking along the main entrance is 269.5° (Magli 2017a, b).

Fig. 11.7 The Bayon temple, Angkor ▶

As a consequence, at sunset on the equinoxes, the setting sun fades into the central tower and the profiles of the faces gradually merge in the glare of the sunlight, producing also a spectacular hierophany (Figs. 11.7 and 11.8).

11.4   Perennial Hierophanies in the Khmer Heartland

The equinoctial hierophanies do not fill the list of all those visible in the Angkor heartland. Indeed, a second, very interesting point is the existence of what can be called perennial hierophanies. Their origin is due to the fact that certain kings, in building their own temples, deliberately created visual connections with already existing monuments of previous kings, in order to establish an explicit reference to specific, dynastic traditions. Such alignments therefore are analogous to those existing in Egypt—think for instance to the "Giza axis".

The existence of topographical alignments between monuments of Angkor was first noticed by Paris (1941) who registered scores of them. Several alignments certainly occur by chance, while others were probably only functional, used to facilitate surveys. After an accurate scrutiny (Magli 2017a, b) there remains, however, a certain number of symbolic, intended alignments. Most of these topographical relationships run along cardinal directions and therefore astronomy was probably of help in realizing them. Unfortunately, today the jungle—which was completely absent at Khmer times—limits the view between the temples and only a

Fig. 11.8 Sunset at the Bayon at the Spring equinox, as seen from the main (east) entrance. The sun merges into the central tower

few of these relationships can be verified by direct sight. We shall limit ourselves here to discuss only two fascinating examples, related to the Mebons.

As mentioned, the Mebons are temples built on artificial islands located in the middle of the barays. These islands were built with the only scope of building the temples over them; most were planned and constructed when their barays were already existing and therefore facing additional technical difficulties. The first example is connected with the very first state temple built in Angkor, the Bakong, built around 880 AD by Indravarman I. The temple was built along with a huge baray, the so-called Indrataka, which was located in such a way that the middle axis lies on the same meridian of the temple, 1.8 km to the south of the baray's center Later, inside the already existing baray, Yasovarman I (successor of Indravarman I) built the Lolei, a Mebon placed along the very same meridian. In this way, a clearly intentional and symbolic alignment was created between the island temple and the state temple of the predecessor.

Around the beginning of the tenth century AD, Yasovarman I initiated the construction of his own enormous artificial lake, the so-called East Baray, which is dried today but originally was a gigantic rectangular basin of 7.5 × 1.8 km oriented to the cardinal points. About 50 years later the king Rajendravarman II decided to pay tribute to this great predecessor with a spectacular architectural project, inspired by the same ideas put forward by Yasorvarman with the construction of Lolei. First, he built his own state temple, Pre Rup, on the same meridian of the center of the East Baray, about 1.3 Kms to the south. Further, he asked to his architects to build an artificial island and a temple over it—today called the East Mebon—at the center of the East baray, and thus almost on the same meridian of Pre Rup. Building such a monument was certainly not an easy task. It is in fact difficult to believe that the island was rapidly assembled during the dry season and it is even more difficult to believe that it was built within the waters, so the enormous reservoir must have been dried voluntarily (by closing the inlet moats) for a suitable amount of time. Its construction certainly expressed a direct will by the king, to implement in a quite spectacular and sophisticated way a dynastic continuity. On the island indeed, a beautiful temple was constructed, called today East Mebon.

These topographical connections are well within the visibility distance between the corresponding monuments and therefore were directly enjoyable at the time of their construction, when the whole area was densely inhabited and cleared so that the towers of the temples were standing out prominently in the landscape. It can thus be said that the builders realized "perennial hierophanies", aimed to recall everyday (also during the night, if we think to fires on the summits of the temples) the greatness of the kings and of their royal ancestors. Still Today, in spite of the presence of the jungle, we can feel the emotion of recognizing the summit of the high towers of East Mebon looking north from the upper terrace of Pre Rup, thus renovating an ideal and spiritual connection devised by a great king more than one thousand years ago (Fig. 11.9).

Fig. 11.9 The towers of the East Mebon temple emerge from the jungle when looking true north from the upper terrace of Pre Rup

11.5   The "Sun Path" at Borobudur

Java is a volcanic island, comparable in size to England. The natural environment of
Java is tropical rain forest, and the humid climate favors the cultivation of rice
which was practiced on large scale since very ancient times, in turn favoring the
establishment of a wealthy kingdom. A special period of flourishing of this kingdom was that of the Sailendra dynasty (8–10 century AD) during which wonderful
temples were constructed. Among them, Borobudur singles out as one of the
greatest Buddhist monuments in the world (Mcguigan 1995). The monument is not
easy to describe; in a nutshell, it is a huge (35 m high, 123 m square base) artificial
hill, composed in sequence by six square platforms and three circular basements
(Fig. 11.10). The circular basements host 72 Buddha statues, each seated inside a
"stupa" (bell-like) chapel. A central, greater stupa crowns the uppermost terrace,
and the walls and balustrades are decorated with fine low reliefs. The temple is
endowed with two satellite temples, Mendut and Pawon, probably built in the same
period.

A number of pages have been written about the symbolism embodied in the
details of the construction and in the numbers devised in the Borobudur architectural features, allegedly connected with astronomical cycles. Also here, as in
Angkor, most of this "numerology" arises by chance and has little or nothing to do
with the true intentions of the builders, which were rather (or mostly) meant as
explicit, easily readable symbols of the religion on which the temporal power of the
kings relied its rights. Nonetheless, astronomy appears to play a relevant role in the
planning of Borobudur. First of all, the square plan of Borobudur proper is quite
precisely oriented to the cardinal points. Second, the 3 temples of the complex are
located along a straight line: starting from the east we find Mendut, Pawon at 1150

Fig. 11.10   The Borobudur temple

m as the crow flies, and finally, further 1750 m west, Borobudur. The line, looking from Mendut towards Borobudur, bears an azimuth of 263°.

This topographical relationship looks hardly casual, as probably a processional way to Borobudur ran along the other two temples in ancient times (the path is still partly mimicked by the modern road from Borobudur to Pawon up to the river Pogo). The orientation of the axis could be chosen at will by placing the temples in an appropriate way, and the macroscopic deviation from the east-west direction appears rather odd at the first sight, since Borobudur—as mentioned—is oriented to the cardinal points. It is, therefore, worth considering the possibility of an astronomically significant orientation (Long and Voute 2008, Magli 2016). The horizon corresponding to the line of sight towards Borobudur has an altitude of 1° 48' yielding a absolute declination of 7° 8'. This declination is very close to the latitude of the place, which is 7° 36' south and therefore, a very likely interpretation is connected to the sun. The axis indeed points quite precisely to the setting of the sun behind Borobudur on the days of the zenith passages, 11 October and 28 February (the orientation of Pawon and Mendut proper might in turn have been related to the Moon, see Magli 2016 for details). The alignment may correspond to a king's ritual connected with the path of the sun in the sky, as proposed by Moens (1951).

Borobudur has few terms of comparison, but astronomy does play a role in the two most strikingly similar monuments. One, of course, is the Bayon temple discussed previously in this chapter, literally crowded on all sides by stone images of benevolent Buddhas. The other is the so-called 108 stupas monument, located on a hillside directly on the western bank of the Yellow River at Qingtongxia, Ningxia, China. This monument (today heavily restored) is only slightly later than Borobudur: it was constructed during the Western Xia dynasty (1038–1227 AD), as part of a greater Buddhist temple complex. It is composed by 108 stupas of sun-dried mud bricks, arranged in rows disposed in a triangular formation which narrows with height, from 19 stupas on the first row to the uppermost single one, so that a front view of this monument is actually quite reminiscent of one side of Borobudur. The azimuth (of the front, looking out) is 120° which, with an horizon height close to zero, gives a impressive declination of -24° that is, very close to the winter solstice sunrise. In the opposite direction, although the hill besides the monument increases the actual declination with respect to that of the summer solstice sunset, the sun can anyhow be seen to merge into the complex on the day which—at the latitude of the site which is not tropical—is the closest approximation of the zenith passage.

References

Bruun, O. (2008). *An introduction to Feng Shui*. Cambridge: Cambridge University Press.
Didier, J. C. (2009). In and outside the square: The sky and the power of belief in Ancient China and the World, III. *Sino-Platonic Papers, 192*, 1–270.

Fletcher, R., Evans, D., Pottier, C., & Rachna, C. (2015). Angkor Wat: An introduction. *Antiquity,* *89,* 1388–1401.

Jacques, C., & Lafond, P. (2004). L'Empire Khmer. Cités et sanctuaires Vth-XIIIth siècles. Paris, Fayard.

Long, M., & Voute, C. (2008). *Borobudur: Pyramid of the cosmic Buddha.* NY: Printworld.

Magli, G. (2017). *Archaeoastronomy of the sun path at Borobudur.* Pre-print arXiv:1712.06486.

Magli, G. (2017b). Archaeoastronomy in the Khmer heartland. *Studies in Digital Heritage, 1*(1), 1–17.

Magli, G. (2018). Royal mausoleums of the western Han and of the Song Chinese dynasties: A satellite imagery analysis. *Archaeological Research in Asia, 15,* 45–54.

Magli, G. (2019a). The sacred landscape of the "Pyramids" of the Han emperors: A cognitive approach to sustainability. *Sustainability, 11*(3), 789.

Magli, G. (2019b). Astronomy and Feng Shui in the projects of the Tang, Ming and Qing royal mausoleums: A satellite imagery approach. *Archaeological Research in Asia, 17,* 98–108.

Mcguigan, Debra L. (1995). The Borobudur, Central Java ca. 732–910 A.D. *CAANS, 16,* 5–18.

Moens, J. L. (1951). *Barabuḍur, Mendut en Pawon en hun onderlinge samenhang.* TBG 84: 326-432 (Barabudur, Mendut and Pawon and their mutual relationship, M. Long, Trans.).

Needham, J. (1959). *Science and civilisation in China, Vol. 3: Mathematics and the science of the heavens and the earth.* Cambridge: Cambridge University Press.

Needham, J. (1970). *Clerks and craftsmen in China and the West: Lectures and addresses on the history of science and technology.* Cambridge: Cambridge University Press.

Paludan, A. (1981). *The imperial ming tombs.* Yale: Yale University Press.

Paludan, A. (1991). *The Chinese spirit road: The classical tradition of stone tomb statuary.* Yale: Yale University Press.

Pankenier, D. (2009). Locating true North in Ancient China. *Cosmology across Cultures, 409,* 128–137.

Pankenier, D. (2015). *Astrology and cosmology in early China: Conforming earth to heaven by David W. Pankenier.* Cambridge: Cambridge University Press.

Paris, P. (1941). L'importance rituelle du Nord-Est et ses applications en Indochine. *Bulletin de la École française d'Extême-Orient, 41,* 303–334.

Stencel, R., Gifford, F., & Moron, E. (1976). Astronomy and cosmology at Angkor Wat. *Science, 193,* 781–787.

Wu, H. (2010). *The art of the yellow springs: Understanding Chinese tombs.* Honolulu: University of Hawai'i Press.

Zhewen, L. (1993). *China's imperial tombs and mausoleums.* Beijing: Foreign Languages Press.

Exercises

The Sun

1. Draw a picture to prove that the altitude of the celestial pole equals the latitude of the observer.
2. Start a virtual planetarium at sunrise on a random day and then follow the azimuth and the declination of the sun at rising up to a solstice and then back again.
3. Today you are in Rome, at dawn, looking at an azimuth of 96°. You see the sun rising from behind a hill which is 10° high. Establish the day(s).
4. Draw a picture to convince yourself that the presence of a hill at the eastern horizon increases the azimuth of sunrise with respect to that corresponding to a flat horizon. What happens to the azimuth of sunset if another hill is placed at the western horizon?
5. Determine the azimuth of the rising sun at the summer solstice in the following cities: Cairo, Milan, London, Stockholm, Quito, Cusco, Santiago, with a flat horizon. How does this azimuth change if the horizon is 5° high?
6. Determine on how many days the sun casts shadows from the north in the following cities: New York, Lima, Mexico City, Buenos Aires, Quito.
7. Determine the Julian date of the summer solstice in AD 400 and in AD 1400, and the proleptic Julian date of the winter solstice in 1000 BC.
8. At a certain place the latitude is equal to L degrees. In the days of the zenith passages of the sun the declination of the sun is D. How much is L-D?
9. A temple located in the northern hemisphere has azimuth 130°, altitude of horizon is zero. During morning, the facade of the temple is illuminated by the sun everyday of the year. Establish if

   - this is never possible (explain)
   - this is possible only below a certain latitude (find the latitude)
   - this is possible only above a certain latitude (find the latitude).

G. Magli, *Archaeoastronomy*, Undergraduate Lecture Notes in Physics,
https://doi.org/10.1007/978-3-030-45147-9

The Stars

1. Determine azimuth and date of the Heliacal rising of Aldebaran in Cairo. Do the same for Milan, London, Cusco, Rio de Janeiro. Estimate the period of invisibility. Solve the same problem for Canopus and for the Pleiades.
2. Determine the latitude at which the seven stars of the big Dipper cease to be circumpolar in the northern hemisphere today. The same in the seventh century BC. Try to explain why Homer says that the Big Dipper "never bathes in the sea".
3. Determine at which latitude the star alpha of the constellation Centaur is visible today in the northern hemisphere.
4. Try to define the heliacal rising of a zodiacal constellation as a whole. Do you think it is a meaningful concept? Consider the same question for the Pleiades.

Precession

1. You are in Cairo, and see Sirius rising during the night with the azimuth 108° on a flat horizon. Establish which century you are living in.
2. You see the sun at the spring equinox rising close to the star alpha of Aries. Establish which century you are living in.
3. Determine the constellation in the background of the winter solstice sun: today, in the first century BC, in 2500 BC, in 10,000 BC. The same for the spring equinox. Does the result depend on the position of the observer?
4. Does the day of Heliacal rising of a star at a fixed latitude change with precession? Explain your answer fully.
5. Determine the declination of Sirius in 2700 BC. Estimate the date of the Heliacal rising of this star at the latitudes of Cairo and of Aswan at that time.
6. Discuss whether Vega has ever been a "Pole Star". The same for Deneb.
7. Define a concept of "south pole star" and establish if there is any naked-eye visible star in the sky which may become, or may have been, such a star.
8. There is a point in the northern sky which is a true, invariable pole. However, it is not located close to any visible star. Why is there such a point? Where is it? What about the southern hemisphere?

Moon and Venus

1. Start a virtual planetarium in a day of year 2000 and follow the behaviour of the Moon at the horizon for a couple of lunar months. Notice the corresponding declinations and the phases of the Moon. Once you are acquainted with this, analyse the behaviour of the Moon in December and in June 2006.
2. You are asked to define, in analogy with the tropics, the concept of "lunar tropics". Where do you think they should be located? Why? Explain analogies and differences with the solar tropics.
3. At which latitude is the Moon at the major northern standstill circumpolar? Explain the result.

4. Start a virtual planetarium on a random day and search for Venus, establishing in which part of her synodic period is situated. Then follow the behaviour of the planet at intervals of a few days for one entire cycle. Concentrate your analysis close to the days of disappearance and reappearance of the planet.
5. Start a virtual planetarium on November 5, 2005. Verify that Venus is going to attain the minimal declination. When do you think that the maximal declination will be attained? Check your hypothesis.

Statistics

1. In a necropolis located in a flat area where the azimuth of the winter solstice is 120°, the azimuths of 35 tombs' corridors are measured, and 12 of them fall into the sector of azimuths of the sun climbing in the sky.

   - Find the probability p that one azimuth is by chance located in this sector
   - Find the probability that 12 azimuths are by chance located in this sector
   - Find the mean and the standard deviation of the distribution
   - Find the sigma level we are working with. Can we claim that the orientations are definitely astronomical?

2. After a measurement, 21 out of 40 Roman towns based on a orthogonal grid turn out to be orientated within ±2° from the cardinal points. Can we claim that the orientation was deliberate? With what degree of confidence?

Virtual Fieldwork

The exercises below are based on important monuments and sites. Before addressing them, readers should acquaint themselves with the basics on the corresponding historical period.

1. Using Google Earth and a Sky Globe software, analyse:

   - The orientation of the Milan cathedral.
   - The orientation of the Chartres cathedral.
   - The orientation of Notre Dame in Paris.

2. Using Google Earth and a Sky Globe software, analyze:

   - The orientation of the Roman town of Merida, Spain.
   - The orientation of the Roman town of Verona, Italy.
   - The orientation of the Roman town of Timgad, Algeria.

3. Using Google Earth and a Sky Globe software, analyse the orientation of the Parthenon in Athens.

   - Do you think that a solar hypothesis is feasible?
   - Try to formulate a alternative stellar hypothesis.

4. Locate the Nasca drainage plain on Google Earth. Then:

   - Become aware that it is a desert but that nearby zones towards the coast are fertile, so people could walk around. Have a look at the horizon profiles.
   - Locate the modern Pan-American highway and search for zoomorphic geoglyphs around it. Try to locate at least five geoglyphs including the so-called monkey, spider, and cormorant.
   - Try to define a symmetry axis for each of the figures. How many azimuths would you measure in fieldwork? Measure the same azimuths and horizon altitudes with Google Earth and create a virtual fieldwork notebook.
   - Analyse the corresponding sky in the first centuries AD. Would you suggest an astronomical orientation for the figures?

5. Locate the Hopewell earthworks called Great Circle and Octagon of Newark (Licking County, Ohio) on Google Earth (the Octagon is easy to find as it is used as a Golf course). Then:

   - Measure the axis of Octagon-circle earthworks and the sight lines defined by two sides of the Octagon and by one corner-to-corner line point. Try to determine whether the corresponding horizon is/was flat in the first centuries AD. Are there other lines to be measured to avoid selection effects?
   - Determine the azimuths of the sun at the solstices and those of the Moon at the lunar standstills at Newark and compare them with the results above. Do you think that the project of the earthworks was based on the moon positions at the horizon or that your results were obtained by chance?

Printed in the United States
By Bookmasters